Simply Good Worksheets for Physics 2
Problem Sets in Electricity, Magnetism, and Waves

Arnold W. Yanof

August, 2017

Copyright © 2017 by Arnold W. Yanof

All rights reserved.

No part of this book may be reproduced by any means, nor transmitted, nor translated into a machine language or graphical or digital form without the written permission of the author.

To college students and teachers everywhere

Useful Physical Constants

Use the Rounded Values from this table to calculate answers to problems in this workbook

Symbol	constant	Rounded Value	Precise Value
c	speed of light	3.0×10^8	2.99792458×10^8 m/s
k	Coulomb Law constant	9×10^9	8.987551787×10^9 Nm^2/C^2
e	electronic charge	1.6×10^{-19}	$1.6021766208 \times 10^{-19}$ C
m_e	electron mass	9.11×10^{-31}	$9.10938356 \times 10^{-31}$ kg
m_p	proton mass	1.67×10^{-27}	$1.672621898 \times 10^{-19}$ kg
N_A	Avogadro's number	6.02×10^{23}	$6.022140857 \times 10^{23}$
$\mu_0/2\pi$	Magnetic Force constant	2×10^{-7}	2×10^{-7} N/A^2
h	Planck's constant	6.63×10^{-34}	$6.62607004 \times 10^{-34}$ J-s

Contents

1 Coulomb's Law   **1**
  1.1 Force between pairs of charges 1
      1.1.1 Proportional reasoning 2
  1.2 Charges in a row 3
      1.2.1 Opposing forces in a line of charges 4
  1.3 Coulomb's Law components 5
      1.3.1 Coulomb's Law - equilateral triangle 6
  1.4 Practice quiz on Coulomb's law 8
      1.4.1 Quiz Answers 9

2 Voltage, charge, and stored energy   **10**
  2.1 Synchrotron particle energy 13
  2.2 Voltage and energy in a circuit 14
  2.3 Voltage near a point charge 15

3 Electric Current and Power   **16**
  3.1 Current examples 18
  3.2 Power examples 19
  3.3 Practice quiz on charge, current, voltage, power, resistance and energy 20
      3.3.1 Quiz answers 21

4 Electrical Resistance   **22**
  4.1 Series and Parallel Resistance 24
  4.2 Dividing Voltage and Current 27
      4.2.1 Voltage divider 27
      4.2.2 Current divider 28
  4.3 Resistor networks 32

5 Electric Field   **34**
  5.1 E-field of a Point Charge 36
  5.2 E-field direction 37
  5.3 E-field components 38
  5.4 Dipole field 43
  5.5 Resistivity 44

6 The Capacitor   **46**
  6.1 Proportional reasoning: capacitance 49

6.2	E-field and Voltage	50
6.3	Dielectrics	51
6.4	Quiz on capacitors and uniform E-field	52
	6.4.1 Answers to capacitor & E-field quiz	53
6.5	Breakdown Field	54
6.6	Capacitors in series and parallel	55
6.7	Capacitor energy	57
6.8	Capacitor circuits	58

7 RC Circuits — 60

8 Magnetic Force Between Currents — 62

	8.0.1 More wires	64
8.1	Force of a current on a charge moving parallel	65
	8.1.1 positive charge	65
	8.1.2 negative charge	65
8.2	B-field around a wire	67
	8.2.1 More B-field around parallel wires	73
8.3	Quiz on force between wires & B-field around a wire	75
8.4	Answers to quiz on force between wires & B-field around a wire	76
8.5	B-field around a current loop	77

9 B-field Force on a Current — 78

9.1	Magnetic force magnitude	79
	9.1.1 More magnetic direction	81
9.2	Net force on a current loop	83
9.3	B-force on a moving charged particle	85
9.4	B-force on a current loop	87
9.5	Torque on a current loop	88
9.6	Quiz on B-field force and torque	90
	9.6.1 Answers to Quiz on B-field force and torque	91

10 The Solenoid — 92

10.1	Solenoid B-field force	94
10.2	Solenoid proportional reasoning	95

11 Faraday's Law and Lenz's Law — 97

11.1	Faraday Applications	100
11.2	Motor / Generator	101
	11.2.1 Rail Gun Motor/Generator	102

12 AC Circuits — 103

13 The Transformer — 106

13.1	power transmission	110
13.2	Quiz on AC Transformers	111
	13.2.1 Answers to Quiz on AC Transformers	112

14 Traveling Waves — 113

14.1 Electromagnetic waves and photons 115
14.2 Light sources - Inverse square law 116
      14.2.1 Light intensity 118
14.3 Refraction and Snell's Law 120

15 Wave interference   126
15.1 Wave Interference from 2 coherent sources 129
15.2 Grating diffraction of light 131
      15.2.1 Diffraction in nature 134

16 Resonance   136
16.1 Sound resonances 138
16.2 Other resonance examples 141

17 Lens optics   144
17.1 Focal length, source and image distances 146
17.2 Ray tracing and image size 148
17.3 Magnified virtual images 153

PREFACE

This workbook is a companion volume to **Simply Good Physics 2: Electricity, Magnetism, and Waves**, by the same author. Together they make a complete single semester course in college physics without calculus. This offers the student and the college teacher a low-cost, light-weight alternative to the bloated offerings of the big publishing houses.

Yet this workbook is complete. It covers all the ideas of electricity, magnetism, and lightwave phenomena. The emphasis is on the basic concepts and formulas. Each chapter and section starts with a brief re-iteration of the variables, equations, and principles which will be explored in the next several worksheets.

These problems have been tested on many years of classes of community college students. The workbook is tutorial, building up to more complicated questions by leading the student through a page or so of intermediate calculations. Answers appear in an upside-down footnote at the bottom of each worksheet. The workbook also provides some sample quizzes, so students have an idea of what professors' expectations will typically be in assessment.

Acknowledgement: This workbook is based upon a methodology of teaching physics developed by Professor Terryl Fender of South Mountain Community College in Phoenix, Arizona. The concept is that problem-solving forms the heart of any physics course. Problems are grouped in worksheets to illustrate basic relationships, dependencies and more advanced concepts. Typically, several worksheets will be done in class with some interactive leadership from the teacher followed by supervision of group problem-solving activity. Then a few worksheets will be assigned for homework, and checked for completion and any questions in the following class.

Each worksheet contains a sequence of questions which lead the student, step-by-step, to the answer. Many of the worksheets in this volume were contributed or adapted from those created by Prof. Fender. Almost all of them have been tested on at least a few classes to ensure that there are a minimum of ambiguities and pitfalls.

Chapter 1

Coulomb's Law

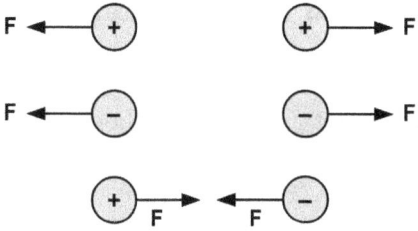

Unlike charges attract; like charges repel

1.1 Force between pairs of charges

$$F = k\frac{Q_1 Q_2}{R^2}, \tag{1.1}$$

the force between two charges Q_1 and Q_2, $k = 9 \times 10^9$, R the distance between them.

Coulomb's Law equation uses the absolute values of the charges to find the magnitude of the force. The direction of the force vector can be specified by the words, *attractive* or *repulsive*, by directions on the paper, or by coordinate directions if they are given.

1. 2 charges A and B each +2 µC are separated by 3 cm. What is the force between them? [1 µC = 10^{-6} C, "one micro-Coulomb"]

2. Two charges that are 1 C are separated by 10 meter. What is the force between them?

3. Two charges are separated by 6 cm. If one charge is 120 nC and the other is -50 nC, what is the force between them?

Answers: 1. 40 N 2. 9×10^7 N 3. 1.5×10^{-2} N

1.1.1 Proportional reasoning

4. The force between two charges is 120 N. If the separation between the charges is increased by a factor of 3, what then is the force between them?

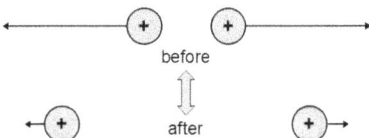

Inverse square law force

5. The force between two charges separated by an unknown distance is 120 N. If the size of each charge is doubled, but the distance remains the same, what then is the force between them?

6. Two unknown, but equal, charges Q are separated by 5 cm. The force between them is 230 N. What is the value of Q?

1.2 Charges in a row

When there are more than two charges, one must calculate the force acting between every pair of charges. Then the net force on any one charge is the sum of the forces due to each of the other charges. The 'sum' in this case means the vector sum.

7. Three charges are lined up in a row as shown [1 nC = 10^{-9} C, "one nano-Coulomb"]:

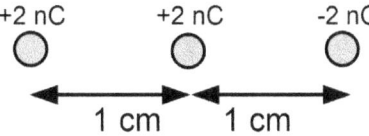

a) What is the force on the middle charge?

b) What is the force on the left charge?

c) In part b), does the middle charge 'block' the effect of the -2 nC charge?

Answers: 7a. 7.2×10^{-4} N   7b. 2.7×10^{-4} N

1.2.1 Opposing forces in a line of charges

In this problem two like charges are placed symmetrically on either side of a middle charge. Each exerts the same magnitude force on the middle charge. But the net force due to the two symmetrically placed charges cancels out to zero net force.

8. Five charges are lined up in a row as shown:

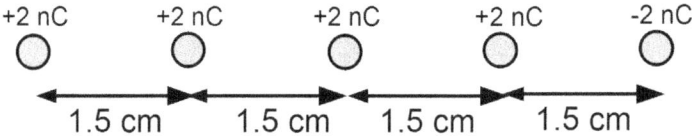

What is the force on the middle charge?

Answers: 8. 8.0×10^{-5} N

1.3 Coulomb's Law components

1. A 1.0 C charge is 5.0 m away from a 2.0 C charge.

a. Draw an arrow from the 2.0 C charge to show the force on the 2.0 C charge, and label it showing the magnitude of the force.

1.0 C   2.0 C

b. Draw an arrow to show the force on the 1.0 C charge, and label it with the value of the magnitude. Which Newton's Law is illustrated here?

2. The 2.0 C charge is now 37° above the horizontal, as shown:

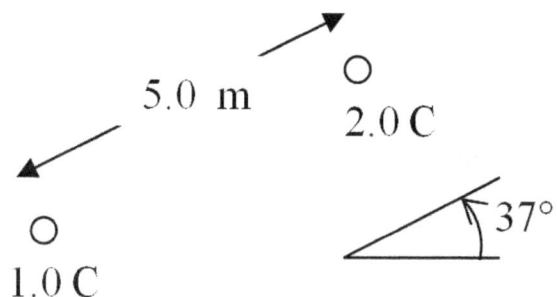

a. Draw arrows to show the x- and y- components of force on the 2.0 C charge.

b. Calculate the x- and y- components:

$F_x =$

$F_y =$

Answers: 1b. $|F| = 7.2 \times 10^8$ N   2b. $F_x = 5.75 \times 10^8$ N $F_y = 4.33 \times 10^8$ N

1.3.1 Coulomb's Law - equilateral triangle

1. Three 2.0 μC charges are arranged in an equilateral triangle as shown. The side of the triangle is 3.0 cm.

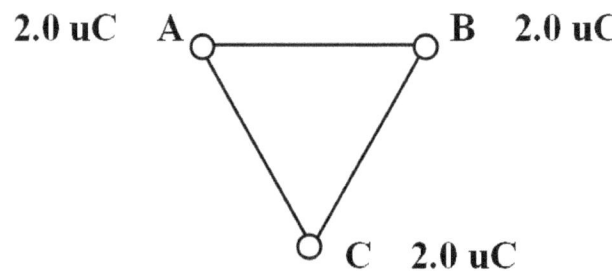

The net force acting on charge C can be found by separating the two forces that act on charge C into x- and y- components.

a. Draw an arrow on charge C showing the force of charge A on charge C. Find the magnitude of this force.

$|F_{A \to C}| =$

b. Calculate the x- and y- components of $F_{A \to C}$:

$F_{A \to C,\, x} =$ $\qquad$ $F_{A \to C,\, y} =$

c. Similarly, calculate the x- and y- components of $F_{B \to C}$:

$F_{B \to C,\, x} =$ $\qquad$ $F_{B \to C,\, y} =$

d. Find the net force on charge C by adding the effects of A and B on C:

$F_x = F_{A \to C,\, x} + F_{B \to C,\, x} =$

$F_y = F_{A \to C,\, y} + F_{B \to C,\, y} =$

Repulsive forces between every pair of like charges

Answers: 1b. 20 N, −34.6 N   1c. −20 N, −34.6 N   1d. −69.3 N

2. Two 2.0 μC charges and a − 2.0 μC charge are arranged in an equilateral triangle as shown. The side of the triangle is 6.0 cm.

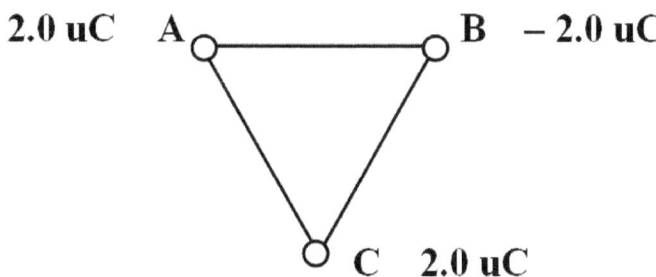

a. Draw arrows on charge C showing the forces due to charge A and charge B. Find the magnitude of this force.

$|F_{A \to C}| = |F_{B \to C}| =$

b. Calculate the x- and y- components of $F_{A \to C}$ and $F_{B \to C}$:

$F_{A \to C, x} =$ $\qquad$ $F_{A \to C, y} =$

$F_{B \to C, x} =$ $\qquad$ $F_{B \to C, y} =$

c. Find the net force on charge C by adding the effects of A and B on C:

$F_x = F_{A \to C, x} + F_{B \to C, x} =$

$F_y = F_{A \to C, y} + F_{B \to C, y} =$

1.4 Practice quiz on Coulomb's law

a. Draw an arrow to show the net force F on the 40 μC charge.

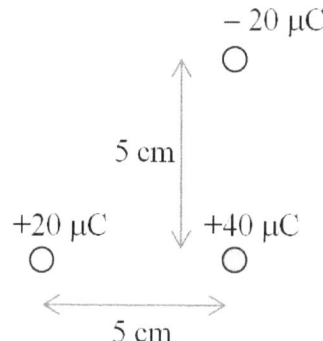

b. Find the x- and y- components and the magnitude of the net force on the 40 μC charge.

c. Find the direction of the force on the 40 μC charge (use $\theta = \arctan(F_y/F_x)$)

Answers: See next page

1.4.1 Quiz Answers

a. The negative charge pulls upward on the 40 μC charge, and the +20 μC pushes the 40 μC charge toward the right. The net force is up and to the right.

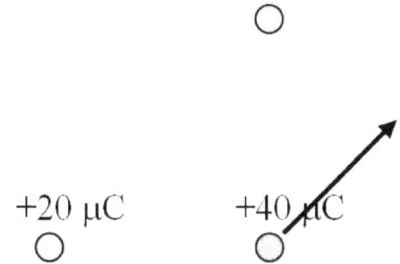

b. The x-component of net force on the 40 μC charge is

$$F_x = |F_{20 \to 40}| = 9 \times 10^9 * \frac{40 \times 10^{-6} \text{ C} * 20 \times 10^{-6} \text{ C}}{0.05^2 \text{ m}^2} = 2880 \text{ N}$$

c. The y-component of net force on the 40 μC charge is numerically the same as the x-component, so $F_y = F_x = 2880$ N. Calculate F_y using the absolute values of the charges. The direction of F_y is upwards because the 40 μC charge and the -20 μC charge have opposite signs.

The magnitude of the net force is given by the Pythagorean Theorem:

$$|F| = \left(F_x^2 + F_y^2\right)^{1/2} = 4073 \text{ N}$$

d. To find the direction of the net force, draw the x- and y- components and draw the resultant. Define the direction by the angle θ of the resultant above the horizontal. Using trigonometry,

$$\tan \theta = F_y/F_x$$

Take the arctangent of both sides to get

$$\theta = \arctan\left(F_y/F_x\right)$$

Chapter 2

Voltage, charge, and stored energy

When we move a positive charge q towards another positive charge, we must do work. Energy is stored in the two-charge system. The potential energy of charge q has increased by an amount

$$\Delta PE = q * \Delta V \tag{2.1}$$

where $\Delta V = V_{end} - V_{start}$ is the change in *voltage* from the starting position to the ending position.

If the charge q we are moving is made twice as large, 2q, but the other positive charge remains the same, then the amount of work is twice as big

$$\Delta PE = 2q * \Delta V \tag{2.2}$$

The change in voltage is still the same.

If the charge q we are moving is changed into a negative charge, $-|q|$, but the other charge remains the same positive charge, then the amount of work is negative, i.e., we have to remove energy from the system. This is because we are holding back on the negative charge as we move it closer to the fixed positive charge.

$$\Delta PE = -|q| * \Delta V \tag{2.3}$$

The change in voltage is determined by the fixed positive charge, and it remains the same ΔV. The change in voltage is called the *potential difference*, and is measured in volts (V).

1. A 1.0 C charge is taken from a point in space with a potential of 10 V to a new place at a potential of 30 V. What is the change in potential energy of the charge?

2. Same 1.0 C charged object starts from rest and moves back from the place at 30 V to the location which has a voltage of 10 V. There is no friction or other force acting against the object.

What is the kinetic energy of the object when it reaches the 10 V location?

3. In a van der Graaf machine, which is a static electricity generator, a spark containing a charge of -500 nC jumps off a terminal whose voltage is - 70,000 Volts. What is the potential energy change of the charge?

4. A charge Q = +4 C passes in a lightning bolt from the positively charged upper thunder cloud to the negatively charged lower part. The energy dissipated is 6.0 Megajoule. What is the voltage difference from beginning to end of the lightning bolt?

Answers: 1. 20 J 2. 20 J 3. -35 mJ (millijoule) 4. -1.5 MV (Megavolt)

5. When a free electron falls into orbit around a free proton to form a hydrogen atom, the electron moves closer to the proton. The voltage change experienced by the electron is $|\Delta V| = 13.6$ Volts. Energy is converted from potential energy into ultraviolet light.

a. As the electron gets closer to the proton, it experiences a   positive   negative   change in voltage (circle one and explain why?)

b. As the electron gets closer to the proton, it experiences a positive   negative change in potential energy (circle one and explain why?)

c. What is the amount of energy in emitted UV light when the electron and proton join to form a hydrogen atom? (the electronic charge is $e = 1.6 \times 10^{-19}$ C)

Answer: 5c. 2.18×10^{-18} J

2.1 Synchrotron particle energy

1. Each time an electron circulates around the ring of an electron synchrotron (atom smasher), it passes through a voltage step and it gains kinetic energy 1.12×10^{-13} J. If the voltage step starts at a voltage of V = 0, what voltage V must it reach in order to pick up this gain in energy?

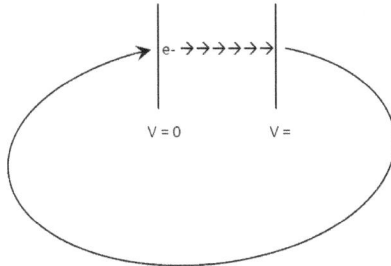

Figure 2.1: Schematic of electron circulating in a synchrotron

2. It takes about 200 C of charge from a 12 Volt battery to start a car.

a. How much energy is required to start the car?

b. Do electrons flow from the   negative   positive   terminal (circle one), through the starting motor, and back to the battery?

Answers: 1. 700,000 V   2a. 2400 J

2.2 Voltage and energy in a circuit

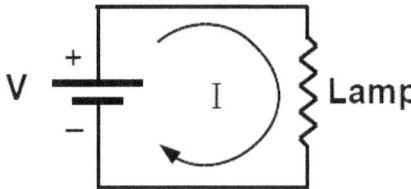

The figure shows a battery lighting a light bulb.

1. a. Suppose V = 1.5 volts, and 6 C of charge goes through the battery. When the charge flows through the battery from the negative terminal to the positive terminal, how much PE does the charge gain?

b. When 6 C of charge goes through the lamp, how much energy does it lose? Where does the energy go?

2. Referring to the same figure, now suppose V = 12.0 V, and 4.0 C of charge passes through the lamp.

a. How much energy is transferred from the battery to the lamp by this movement of charge?

b. If the motion of charge occurs in 3.0 seconds, what is the power in Watts (rate of energy transfer)?

Answers: 1a. 9 J   1b. 9 J   2a. 48 J   2b. 16 W

2.3 Voltage near a point charge

When (+) charge q approaches another (+) charge Q, the potential energy increases. The change in potential energy of charge q is $\Delta PE = q(V_f - V_i)$. Suppose charge q moves from infinitely far away, where the PE = 0 and $V_i = 0$, to a point that is a distance r from Q. Then the PE for q at r is given by the formula, $PE(r) = kQq/r^2$. The voltage at point r due to Q is

$$V(r) = \frac{PE}{q} = \frac{kQ}{r} \qquad (2.4)$$

The voltage in the vicinity of several charges, Q_1, Q_2, ... is

$$V(r) = \frac{kQ_1}{r_1} + \frac{kQ_2}{r_2} + ... \qquad (2.5)$$

where r_1 is the distance between point r and charge Q_1, r_2 is the distance between point r and charge q_2, etc.

1. a. What is the voltage 1 cm away from a charge of +1.0 nC?

   b. What is the voltage 1 cm away from a charge of −1.0 nC?

2. What is the voltage at the center of the equilateral triangle in section 1.3.1 problem #1? Hint: the distance from the center of the triangle to each charge is 1.73 cm.

3. What is the voltage at the center of the equilateral triangle in section 1.3.1 problem #2?

Answers: 1. 900 V   2. 3.12×10^6 V   3. 1.04×10^6 V

Chapter 3

Electric Current and Power

When charge flows continuously from one place to another, that is called an *electric current*. The electric current is the rate of charge transfer, measured in Coulombs per second. The unit of current is the ampere, symbolized **A**. 1 A = 1 Coulomb/second.

Written in symbols, we can find the current from the equation,

$$I = \frac{q}{t} \tag{3.1}$$

where I is in amperes, q in Coulombs, and t in seconds.

Current is mainly discussed in electric circuits, where the charge moves through wires, batteries, lamps, and other circuit elements.

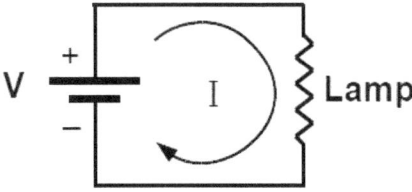

Figure 3.1: Battery lighting a light bulb

1. In the figure, V = 12.0 V, and 4.0 C of charge passes through the lamp in 3.0 s.

a. How much current is flowing?

b. By definition, one can calculate the power from energy and time:

$$Power = energy/time$$
$$P = (qV)/t \qquad (3.2)$$

This equation can be rearranged to give

$$P = V\frac{q}{t} = VI \qquad (3.3)$$

c. Calculate the power in this example using the equation, P = I V

Answers: 1a. 1.33 A   1c. 16 W

3.1 Current examples

1. In a van der Graaf spark 550 nC jumps to my finger in 110×10^{-6} sec. What is the electric current?

2. A charge Q = +6 C passes in a lightning bolt from the positively charged upper thunder cloud to the negatively charged lower part. The time duration of the bolt is 0.2 msec. What is the electric current? 1 millisecond = 10^{-3} s.

3. An electron travels once around a hydrogen atom in 2.5×10^{-17} s. What is the electric current carried by the electron? (The electronic charge is 1.6×10^{-19} C)

4. If one electron from each atom in 63 grams of copper wire moves through the wire in 1 second, how much current would that be? [Hint: 63 grams is one mole of copper]

Answers: 1. 5×10^{-3} A 2. 3×10^{4} A 3. 6.4×10^{-3} A 4. 96,400 A

3.2 Power examples

Referring to Equation (3.3), P = I * V is the power in a circuit that has current I and voltage V. Use this equation to solve the following problems.

1. In your house, the effective voltage is 110 volts. If a light bulb is rated at 100 Watts, what is the electric current in the light bulb?

2. Your hair dryer operates at 1875 W. If the voltage is 110 V, what is the electric current?

3. My January bill for 30 day's worth of electricity shows we consumed 2250 kWh of electricity. Assuming the effective voltage is 110 Volts, what was the average electrical current into my house in January?

4. An electric car uses 12,000 W of electric power to cruise at 60 mph. If the car battery is 216 V, what is the current in the electric motor?

Answers: 1. 0.91 A   2. 17.0 A   3. 28.4 A   4. 55.6 A

3.3 Practice quiz on charge, current, voltage, power, resistance and energy

Question 1. A battery powers a night light in my bedroom. The battery voltage is 5.85 V and the current through the night light is 2.37 mA.

a. What power is being used by the night light?

b. The battery stores a charge of 39000 C. How many seconds will the night light work before the battery runs out?

c. How much energy does the battery have stored in it when new?

d. What is the resistance of the night light?

3.3.1 Quiz answers

a. P = I * V = 2.37×10^{-3} A $\times$ 5.85V = 0.0139 W.

b. t = Q/I = 39000 C / .00237 A = 1.65×10^7 s.

c. PE = Q * V = 39000C $\times$ 5.85 V = 2.28×10^5 joules.

d. R = V/I = 5.85 V / .00237 A = 2.47×10^3 Ω.

Chapter 4

Electrical Resistance

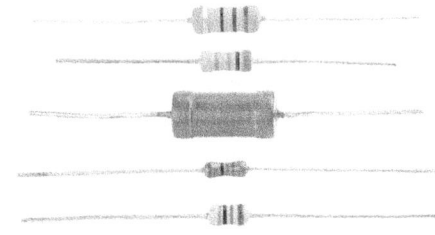

Larger resistors handle more power. Colored stripes encode resistance values.

In an electrical circuit, the current almost always increases when the voltage increases. For many types of loads, there is an almost exact proportionality between voltage and current. The constant of proportionality is called the *resistance*. The unit of resistance is the *Ohm*, symbolized by Greek letter, Ω.

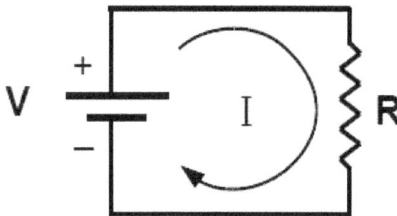

Figure 4.1: Battery pushing current through an arbitrary load resistance, R

In the figure, the zig-zag line symbolizes a load resistance. The value of resistance is

$$R = \frac{V}{I} \tag{4.1}$$

If V is given in volts, and I is given in amperes, then equation (4.1) gives the resistance, R, in ohms. The proportionality in equation 4.1 is called *Ohm's Law*.

1. An old-style tungsten flashlight bulb allows 0.5 A to flow when attached to a 3.0 V battery. a. What is the resistance of the light bulb? b. What is the power when the light is lighted?

Answer: 1a. 6.0 Ω 1b. 1.5 W

2. An LED flashlight operates at 0.01 A when connected to a 2.0 V source of current. What is the resistance of the LED, and what is the power?

3. A Cell phone fully-charged holds 1200 C. It can go 72 hours between charges if there are no calls. a. What is the current drain with no calls? b. If the battery has a voltage of 2.5 V, what is the resistance of the cellphone allowing this current drain, and how much power is being drawn?

4. An electric car uses 12,000 W of electric power to cruise at 60 mph. a. If the car battery is 216 V, what is the current in the electric motor? b. What is the load resistance of the motor?

5. The electrical resistance of dry skin is about 200 KΩ. (1 KΩ = 1000 Ω) If you grab the positive and negative terminals of a 6.0 V battery, how much current flows into your body?

Answer: 2a. 200 Ω, 0.02 W   3a. 4.63×10^{-3} A   3b. 540 Ω, 1.16×10^{-2} W   4a. 55.6 A   4b. 3.88 Ω   5. 3×10^{-5} A

4.1 Series and Parallel Resistance

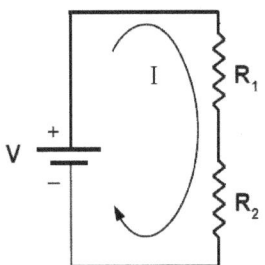

Figure 4.2: Series resistors

When multiple loads are arranged in a daisy-chain fashion, this is called a *series* circuit. The same identical current must flow through every element of the circuit. When two loads, R_1 and R_2 make up the series circuit, the total load resistance is

$$R_T = R_1 + R_2 \qquad (4.2)$$

When loads are arranged like rungs on a ladder, this is called a *parallel* circuit. Each load in a parallel circuit is subject to the same voltage.

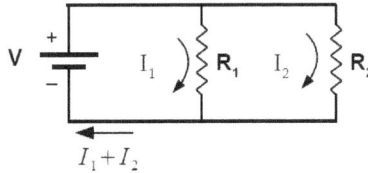

Figure 4.3: Parallel resistors

When two loads, R_1 and R_2 make up a parallel circuit, the total load resistance is

$$\frac{1}{R_T} = \frac{1}{R_1} + \frac{1}{R_2} \qquad (4.3)$$

1. Two resistors, each having resistance 4 Ω, are connected in series. a. What is the total resistance? b. If the two resistors are connected in parallel, what is the total resistance?

Answer: 1a. 8 Ω 1b. 2 Ω

4.1. SERIES AND PARALLEL RESISTANCE

2. Are the different appliances in your home connected in series or in parallel? Explain why.

3. An electric heating coil operating on 100 V has two equal segments, each 1.0 Ω. The 'low' setting for the heater puts the voltage across the two segments connected in series. The 'medium' setting connects just one of the segments. The 'high' setting connects the two segments in parallel. How much is low, medium and high power, in Watts?

4. What is the combined resistance of five 10 Ω resistors connected in parallel?

5. We determine that an existing circuit uses a resistor of 50 Ω, but the circuit would perform better if the value were 40 Ω. What value of resistor could we put in parallel with the 50 Ω resistor, so that the combined resistance would be 40 Ω?

Answers: 3. 5000 W, 10,000 W, 20,000 W   4. 2 Ω   5. 200 Ω

6. Three resistors are provided: 6 Ω, 12 Ω, and 24 Ω. What are all the possible resistances obtained when these three are hooked up in series, in parallel, and in series/parallel?

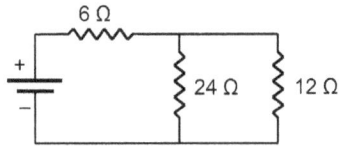

One possible series/parallel circuit

7. My home consumes power at an average rate of 1200 W, night and day. If the voltage from the power company is 240 V, what is the average current consumed? What is the average resistance of my home?

8. A 10 Ω resistor is in series with a 500 Ω resistor. What is the combined resistance? If the two resistors are connected in parallel, what is the combined resistance?

9. When talking on my cellphone, the power dissipated is 5.0 W. When I am talking on the cellphone and using the camera at the same time, the power is 6.0 W. Since these two functions are in parallel, what is the resistance of the 'talking' circuit, of the two circuits together, and of the camera circuit? Assume the lithium battery power is 3.2 Volt.? (Hint: first find the current used by the 'talking' circuit, and the current used by both the talking + camera circuits together)

Answers: 6. series: 42 Ω, parallel: 3.43 Ω, series-parallel: 28 Ω, 14 Ω, 16.8 Ω   7. 5.0 A, 48 Ω   8. 510 Ω, 9.80 Ω   9. $R_{talk} = 2.05$ Ω, $R_{camera} = 10.31$ Ω, $R_{both} = 1.71$ Ω

4.2 Dividing Voltage and Current

In the series circuit of Figure 4.2, each charge q gets energy qV from the battery, and this energy is partly lost in R_1 and partly lost in R_2. Therefore we say the voltage is *divided* by R_1 and R_2 and that R_1, R_2 are a voltage divider.

Similarly, in the parallel circuit of Figure 4.3, the current coming from the battery divides between the R_1 and R_2 branches. In a parallel circuit, the resistors divide the <u>current</u>.

4.2.1 Voltage divider

1. A 120 V power source is connected to two resistors as shown in the figure. a. What is the voltage at points A, B, and C? b. What is the power in each resistor?

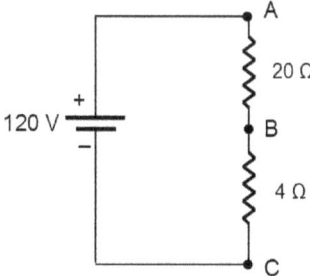

Hint: Find the total resistance, and use this to find the current in the circuit. Then use V = I*R for each of the resistors separately to find the amount of voltage change due to each resistor.

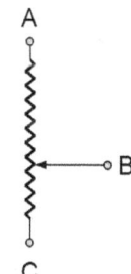

Symbol for a ***potentiometer*** – a variable voltage divider. B is attached to a slider on the resistor. Full voltage is applied between A and C, and reduced voltage is obtained between B and C.

Answers: 1a. $V_A = 120$ V, $V_B = 20$ V, $V_C = 0$ V   1b. $P_{20\Omega} = 500$ W, $P_{4\Omega} = 100$ W

4.2.2 Current divider

1. A 24 V power source is connected to two resistors as shown in the figure. a. What is the total current, I_T? b. What are the currents in each of the two resistors?

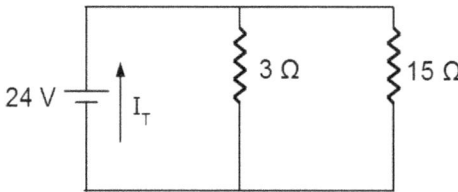

Answers: 1a. $I_T = 9.6$ A   1b. $I_{3\Omega} = 8$ A, $I_{15\Omega} = 1.6$ A

4.2. DIVIDING VOLTAGE AND CURRENT

RIVP method

The 'RIVP' method is an orderly way of calculating and presenting the current, voltage, and power of every resistor in a circuit. Here is a typical use of the method:

Three resistors are combined in series-parallel as shown in the figure. Use RIVP to analyze all the currents, voltages and power.

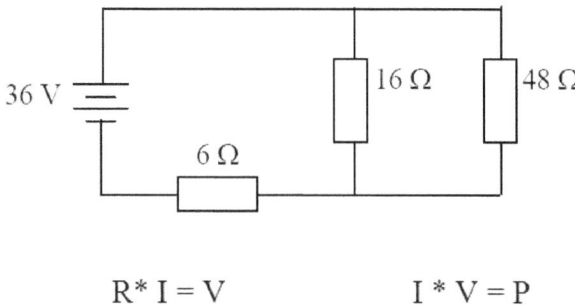

R * I = V   I * V = P

	R Ohms	I Amps	V Volts	P Watts
48 Ω	48	0.5	24	12
16 Ω	16	1.5	24	36
$R_\|$	12	2	24	48
6 Ω	6	2	12	24
R_T	18	2	36	72

Procedure:

1. First step is to put in the known quantities, i.e., 48, 16, and 6 in the 'R Ohms' column, and 36 in the 'V Volts' column.

2. Calculate the parallel resistance value and enter $R_\| = 12 \ \Omega$ in the R Ohms column.

3. Calculate the total resistance and enter $R_T = 18 \ \Omega$.

4. Find the total current and enter $I_T = 2$ A in the 'I Amps' column.

5. The same value of $I_T = 2$ A is entered for the current in the 6 Ω resistor and in $R_\|$.

6. Using R*I = V, enter the voltages in the 6 Ω resistor and in $R_\|$, 12 V and 24 V, resp.

7. All the resistors in a parallel combination receive the same voltage. Therefore the voltage on the 48 Ω and 16 Ω resistors is the same as the voltage across $R_\|$.

8. Using I = V/R, enter the currents 0.5 A and 1.5 A for the 48 Ω and 16 Ω resistors, resp.

9. Finally use I*V = P to enter the entire 'P Watts' column.

1. Use RIVP to complete the table for the circuit below.

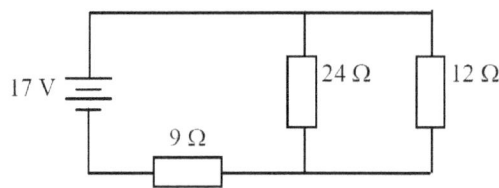

	R Ohms	I Amps	V Volts	P Watts
24Ω				
12 Ω				
R∥				
9 Ω				
R_T				

2. Use RIVP to fill in the table below.

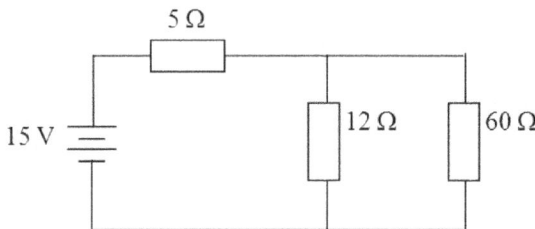

	R Ohms	I Amps	V Volts	P Watts
5 Ω				
60 Ω				
12 Ω				
R∥				
R_T				

4.2. DIVIDING VOLTAGE AND CURRENT

3. Four resistors are combined in series-parallel.

a. Use RIVP to analyze all the currents, voltages and power.

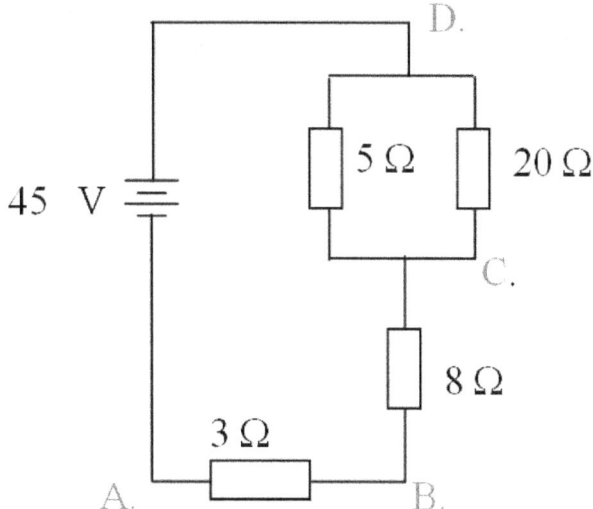

item	R Ohms	I Amps	V Volts	P Watts
R_T				

b. Which nodes (A, B, C, D) would you connect to a voltmeter to obtain a reading of 24 Volts?

4.3 Resistor networks

In more complex resistor networks, look for two resistors in series, or two resistors in parallel. Find the sum of those two resistors, then redraw the circuit with a single resistor replacing the two. Repeat the process until the total circuit resistance becomes obvious.

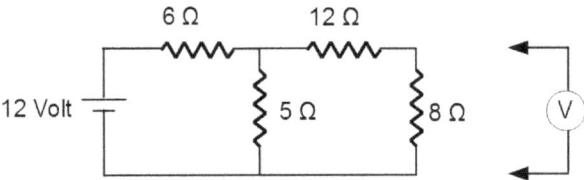

1. In the above circuit the battery produces 12 V. Find

a. Total resistance of the circuit

b. Current through the battery

c. Voltage difference applied across the 5 Ω resistor

d. Voltage read by voltmeter V across the 8 Ω resistor

e. Power produced by the battery

Answers: a. $R_T = 10$ Ω   b. 1.2 A   c. 4.8 V   d. 1.92 V   e. 14.4 W

4.3. RESISTOR NETWORKS

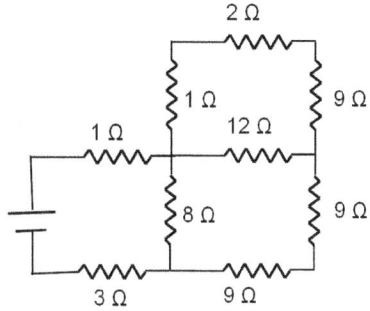

2. Find the total resistance in the circuit as viewed by the battery in the above circuit.

Chapter 5

Electric Field

A point (+) charge produces a radially directed force field

The electric field is a *force* field. Every point in space has a value of the electric field. The electric field is a vector, E. It points in the direction of the force exerted on a charge, q. The magnitude of the force is

$$F = qE \tag{5.1}$$

Because the unit of force is the Newton (N), and the unit of charge is the Coulomb (C), therefore the unit of electric field is N/C, or 'newtons per coulomb'.

<u>Example</u> The electric field at a point in space is 10,000 N/C. If there is a charge q = 35 nC at this point, what is the force on q?

Answer: F = 35×10^{-9} × 10^4 = 3.5 ×10^{-4} N.

1. At a point in space, the electric field E = 2×10^6 N/C. What is the force on an electron (charge = -1.6×10^{-19} C)?

2. An electron has mass 9.11×10^{-31} kg, and charge -1.6×10^{-19} C. An electric field, E, causes it to accelerate at 3×10^7 m/s^2. What is the magnitude of E?

Answers: 1. 3.2×10^{-13} N 2. 1.71×10^{-4} N/C

3. An electric field $E = 1 \times 10^5$ N/C is applied to a charge $Q = 2 \times 10^{-6}$ C. We apply an equal and opposite force on the charge and slowly move it a distance of 3.0 m. What is the potential energy increase of the charge?

4. A charge $Q = 2 \times 10^{-6}$ C is initially at rest. We apply an electric field $E = 1 \times 10^5$ N/C to the charge. We allow it to accelerate over a distance of 3.0 m. E remains constant over this distance. What is the kinetic energy increase of the charge?

Answers: 3. 0.6 J   4. 0.6 J

5.1 E-field of a Point Charge

If a charge Q is located at point A, and there is another charge q located at point B, there will be a force between the charges given by Coulomb's law. Assume the distance between A and B is r. According to (1.1), the force between the two charges is kQq/r^2.

But according to (5.1), the force is also given by qE. Comparing these two statements, the E-field due to Q which acts upon q is

$$E = k\frac{Q}{r^2} \qquad \text{E-field due to point charge} \tag{5.2}$$

1. I am 1.0 meter away from a 1.0 C charge. a. What is the electric field where I am? b. What is the electric field 2.0 meter from the 1.0 C charge?

2. The electric field at a location near a point charge is 120 N/C. If the charge is increased by a factor of 4, what is the E-field at the same location?

3. The electric field near a point charge is 120 N/C. If our distance from the charge is increased by a factor of 3, what then is the electric field?

4. What is the E-field 6 cm to the East of a -50 nC point charge? (Give both magnitude and direction)

Answers: 1a. 9×10^9 N/C   1b. 2.25×10^9 N/C   2. 480 N/C   3. 13.3 N/C   4. 1.25×10^5 N/C, West

5.2 E-field direction

The E-field points *away from* a positive charge, and *towards* a negative charge.

1. Two charges are arranged as shown:

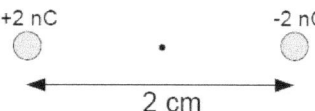

What is the E-field at the point half way in between them?

2. Four charges are arranged as shown below. a. What is the E-field at point **A**? What is its direction?

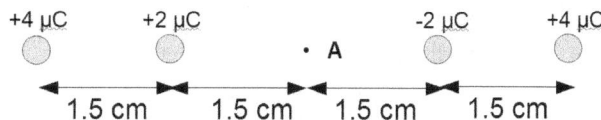

b. What is the E-field experienced by the $-2\ \mu C$ charge, *due to the other three* charges? What is the direction?

Answer: 1. 3.60×10^5 N/C 2. a. 1.60×10^8 N/C, right b. 1.22×10^8 N/C, left

5.3 E-field components

1. A 1.0 C charge is 5.0 m away from a 2.0 C charge. a. Draw an arrow from the 2.0 C charge to show the E-field experienced by the 2.0 C charge, due to the 1.0 C charge; and label it showing the magnitude of the E-field.

$|E_{2\to1}| =$ $\qquad\qquad\qquad\qquad$ $|E_{1\to2}| =$

$\qquad\qquad$ O $\qquad\qquad\qquad\qquad\qquad$ O

$\qquad\qquad$ 1.0 C $\qquad\qquad\qquad\qquad\quad$ 2.0 C

b. Draw an arrow to show the E-field on the 1.0 C charge, and label it with the value of the magnitude. Are the E-fields equal and opposite?

2. The 2.0 C charge is now 37° above the horizontal, as shown:

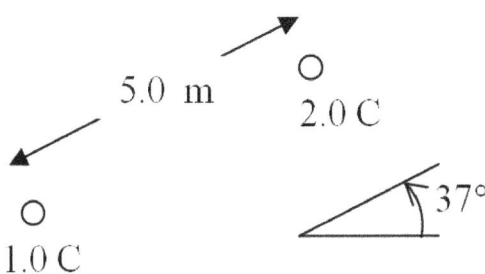

a. Draw arrows to show the x- and y- components of electric field $E_{1\to2}$, experienced by the 2.0 C charge due to the 1.0 C charge.

b. Calculate the x- and y- components of the E-field produced by the 1.0 C charge, acting at the location of the 2.0 C charge (Hint: use $\cos 37°$ and $\sin 37°$):

$E_{1\to2,x} =$

$E_{1\to2,y} =$

Answers: 2b. $E_x = 2.88 \times 10^8$ N/C $E_y = 2.17 \times 10^8$ N/C

5.3. E-FIELD COMPONENTS

c. Find the magnitude of $E_{1\to 2}$, using the Pythagorean theorem.

$|E_{1\to 2}| =$

Answers: 2b. $|E_{1\to 2}| = 3.60 \times 10^8$ N/C N

3. A $+2.0$ μC and a $-2.0 \mu C$ charge and point P are arranged in an equilateral triangle as shown. The side of the triangle is 6.0 cm.

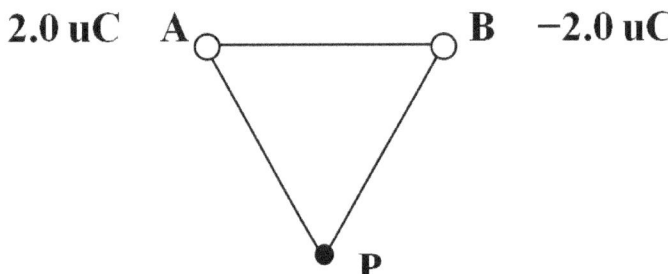

a. Draw an arrow at point P showing the E-field due to charge A at point P. Find the magnitude of this E-field.

$|E_{A \to P}| =$

b. Calculate the x- and y- components of $E_{A \to P}$:

$E_{A \to P, x} =$

$E_{A \to P, y} =$

c. Calculate the x- and y- components of $E_{B \to P}$:

$E_{B \to P, x} =$

$E_{B \to P, y} =$

d. Now sum the x- and y- components separately to get the totals of E_x and E_y at point P. Use the Pythagorean theorem to find $|E|$ at P.

$E_x = E_{A \to P, x} + E_{B \to P, x} =$

$E_y =$

$|E| = \sqrt{(E_x)^2 + (E_y)^2} =$

Answers: 3a. 5.0×10^6 N/C 3b. $E_{A \to P, x} = 2.5 \times 10^6$ N/C $E_{A \to P, y} = -4.33 \times 10^6$ N/C 3d. 5×10^6 N/C

5.3. E-FIELD COMPONENTS

3. Two $+2.0\ \mu C$ charges and point P are arranged in an equilateral triangle as shown. The side of the triangle is **3.0** cm.

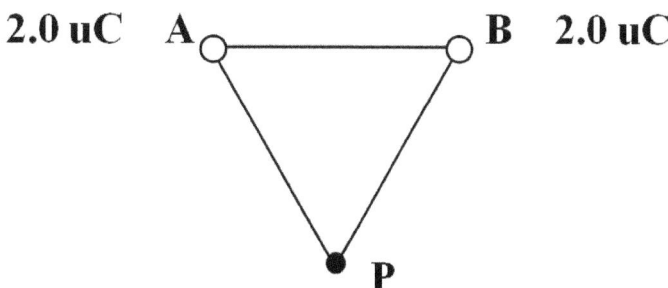

a. Draw an arrow at point P showing the E-field due to charge A at point P. Find the magnitude of this E-field.

$|E_{A \to P}| =$

b. Calculate the x- and y- components of $E_{A \to P}$:

$E_{A \to P,\ x} =$

$E_{A \to P,\ y} =$

c. Calculate the x- and y- components of $E_{B \to P}$:

$E_{B \to P,\ x} =$

$E_{B \to P,\ y} =$

d. Now sum the x- and y- components separately to get the totals of E_x and E_y at point P. Use the Pythagorean theorem to find $|E|$ at P.

$E_x = E_{A \to P,\ x} + E_{B \to P,\ x} =$

$E_y =$

$|E| = \sqrt{(E_x)^2 + (E_y)^2} =$

Answers: 2a. 2×10^7 N/C   2b. 1×10^7 N/C, -1.73×10^7 N/C   2d. $E_x = 0$, $E_y = -3.46 \times 10^7$ N/C

4. Draw an arrow to show the net E-field on the 40 μC charge due to the other two charges.

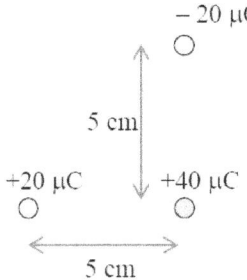

b. Find the x- and y- components and the magnitude of E at the 40 μC charge. (Add components due to the +20 μC and the −20μC)

$E_x =$

$E_y =$

$|E| =$

c. Use $F = qE$ to find the force on the μC charge.

Answers: 4b. $E_x = 7.2 \times 10^7$ N/C $E_y = 7.2 \times 10^7$ N/C

5.4 Dipole field

The figure shows two fixed charges, one positive and one negative. Four points in the space surrounding the charges are labeled **P**, **S**, **T**, and **V**.

a. Consider the E-field due to the positive charge: At each of the four labeled points, draw an arrow indicating the magnitude and direction of the E-field due to the + charge.

b. Now the E-field due to the negative charge: At each of the four labeled points, draw an arrow indicating the magnitude and direction of the E-field due to the − charge.

c. Use graphical vector addition to draw the total E-field at each labeled point due to both charges together.

5.5 Resistivity

Every circuit load has a particular resistance, **R**. Resistive materials can be shaped to have a greater resistance by making them longer and thinner. A given material has an intrinsic constant resistivity, ρ, which is the same no matter what the shape. The equation relating resistance and resistivity is

$$R = \rho \frac{l}{A} \tag{5.3}$$

where l is the length of the sample, and A is the cross-sectional area.

We want units of ρ to be Ω-m, but note that ρ is often given in ohm-cm. For example, the resistivity of copper is 1.68×10^{-8} Ω-m. However, some sources show 1.68×10^{-6} Ω-cm. The conversion is just:

$$[\text{ohm-m}] = [\text{ohm-cm}] \, \frac{1 \; m}{100 \; cm} \tag{5.4}$$

Here is a table of resistivity values for some common metals, semi metals and insulators:

material	resistivity, ρ (Ω-m)
copper	1.68×10^{-8}
silver	1.59×10^{-8}
gold	2.44×10^{-8}
iron	9.71×10^{-8}
graphite	12×10^{-5}
glass	3.84×10^{14}
pure water	1.68×10^5

1. What is the resistance for current flowing between opposite faces of a block of copper that is 1 cm^3?

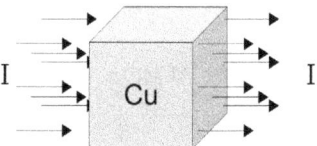

2. What is the resistance of a copper wire that is 75 cm long and .001 cm^2 in cross-section?

Answers: see next page

5.5. RESISTIVITY

3. A wire is 300 m long and has cross-sectional area A = 0.64 mm². It has a resistance R = 8.0 Ω. What is the resistivity?

4. Consider a wire made from the same material as the one in question 3) above. It is 10 times as long and one half the diameter. What is the resistance of this wire?

5. A bar **B** of conducting material has a resistance of 10 Ω. The bar is cut lengthwise in two directions, as shown in the figure; and the four pieces are joined end-to-end. What is the resistance of the new, longer and narrower shape **C**?

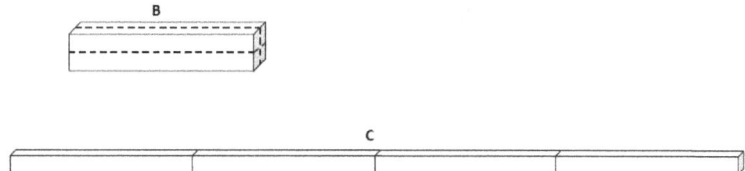

Figure 5.1: A resistance is cut twice and re attached to form a long, narrow resistance.

Answers: 1. 1.68×10^{-6} Ω   2. 126 Ω   3. 1.71×10^{-8} Ω-m   4. 320 Ω   5. 160 Ω

Chapter 6

The Capacitor

Two metal plates separated by a gap of non-conducting material form a useful circuit element called a *capacitor*. The two plates hold separate, equal and opposite charges. This separation of + and − charges means the capacitor can store energy. Also, the E-field flows uniformly out of the + charges and into the − charges. This gives a very simple, uniform E-field configuration, shown in the figure.

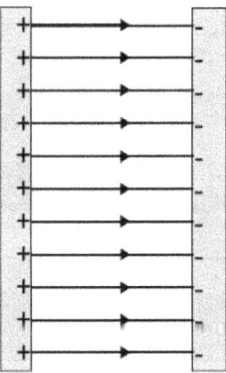

Figure 6.1: Two metal plates with equal and opposite charges form a capacitor.

If a *free* + charge is moved towards the + plate, it experiences an increase in potential energy. Therefore, the voltage of the + plate is positive relative to the − plate. The voltage difference between the plates is proportional to the charge on the plates:

$$V = \left\{\frac{1}{C}\right\} Q \tag{6.1}$$

where Q is absolute value of the charge on either plate, V is the voltage difference between the plates, and 1/C is the constant of proportionality between Q and V. C is called the *capacitance*, measured in farads (F).

If the separation between plates is d, and the area of each plate is A, the capacitance is

$$C = \frac{A}{4\pi k d} \tag{6.2}$$

1. In a 1.0 F capacitor, if the separation between the plates is 1.0 cm, what is the area of the plates?

2. Three parallel metal plates each have area 0.20 m^2. These are stacked horizontally, so that the gap between the middle and bottom plates is d = 0.125 mm, and the gap between the middle and top plates is also 0.125 mm. The top and bottom plates are wired together so as to be at the same voltage V>0, while the middle plate is held at ground potential, V=0.

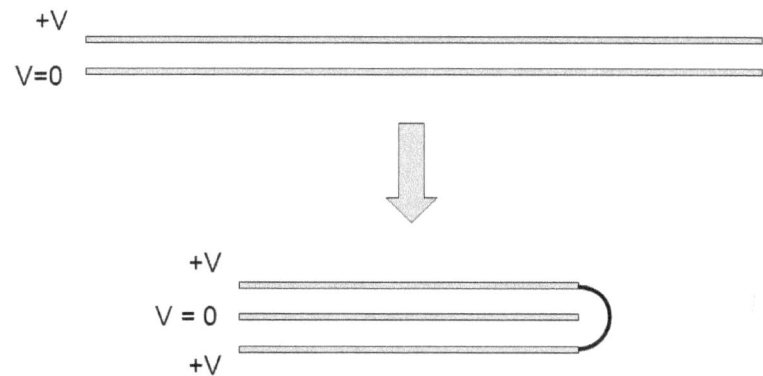

Figure 6.2: A two-plate capacitor is folded over to form an equivalent 3-plate capacitor.

What is the capacitance of the overall structure? (Hint: in effect, this is a capacitor with gap d and area 2A, folded over to form a 3-plate capacitor as shown in the above figure. Hence, the overall capacitance will be twice as big as a parallel plate device with area A and gap d).

Answers: 1. 1.13×10^9 m^2 2. 2.83×10^{-8} F = 0.0283 μF

3. A radio frequency tuning capacitor consists of a comb of 21 plates with each pair having an overlap area of 4.0 cm². The separation between the plates is 1.5 mm. a. What is the capacitance of the tuning capacitor?

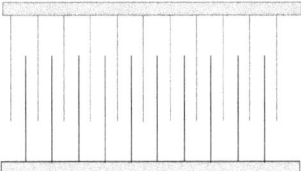

Figure 6.3: A 21-plate capacitor has 11 upper plates and 10 lower plates.

b. If 500 V is applied between the top and bottom combs of plates, what is the stored charge?

4. Two plates having area 0.1 m² are separated 2.0 cm and charged by a van de Graaf generator to a voltage of 60,000 V. What is the stored charge?

5. A 220000 μF capacitor is charged with 0.8 C. What is the voltage on the capacitor?

Answers: 3. a. 4.72×10^{-11} F = 47.2 pF b. 0.0236 μC 4. 2.65 μC 5. 3.64 V

6.1 Proportional reasoning: capacitance

6. A parallel-plate capacitor has an area A = 200 m² and a capacitance C = 0.12 μF. To construct a capacitor with 5× the capacitance, and the same gap distance, what must be the new plate area?

7. A capacitor is constructed from two square, parallel plates, having capacitance C = 210 μF. If a new capacitor has all dimensions multiplied by 3 — the sides of the square plates as well as the gap separating them — what is the new capacitance, C_{new}?

8. A parallel-plate capacitor has a charge Q = 340 nC and a voltage of 125 V between the plates. If the distance between the plates is increased by a factor of 2, what will be the new voltage difference between the plates?

Answers: 6. 1000 m² 7. C_{new} = 630 μF 8. 250 V

6.2 E-field and Voltage

Equation (5.1) shows that an E-field exerts force $F = qE$ on a charge q. If we displace a positive charge *against* the E-field, the PE of the charge increases:

$$\Delta PE = -F\Delta x = -qE\Delta x \tag{6.3}$$

where the − sign occurs because the direction of dispacement is opposite to the direction of E. But the PE change is also given by Equation (2.1) $\Delta PE = q\Delta V$. Therefore the E-field is the negative of the slope of V:

$$E = -\frac{\Delta V}{\Delta x} \tag{6.4}$$

1. A parallel plate capacitor has an gap d = 1.0 mm. A voltage V = 500 V is applied to the capacitor. a. What is the value of E-field between the plates?

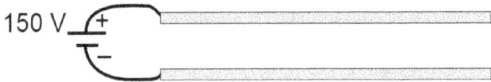

b. In the figure, draw in the E-lines of force, using arrow heads to show the direction of the E-field.

c. If an electron is placed half way between the two plates, what will be its acceleration ($m_e = 9.11 \times 10^{-31}$ kg, $e = 1.6 \times 10^{-19}$ C)

Answers: 1. a. 5×10^5 N/C   c. 8.78×10^{16} m/s^2

6.3 Dielectrics

If a material with dielectric constant κ fills the gap in a capacitor, the capacitance increases by a factor of κ and equation (6.2) becomes

$$C = \frac{\kappa A}{4\pi k d} \qquad (6.5)$$

Here is a table of common dielectrics for capacitors:

material	dielectric constant, κ
vacuum	1
air	1.00059
water	80
glass	7
wax paper	3
mica	5
titanium dioxide	120
mylar	3.1

1. If the capacitor dielectric consists of wax paper that is 10^{-4} m thick, what area is needed to create a capacitance of 1 μF ?

2. A capacitor is made by putting 0.127 mm mylar between a pair of plates having area A = 0.28 m^2. If 1000 V is applied to the capacitor, what is the stored charge?

Answers: 1. 3.77 m^2 2. 6.04×10^{-5} C

6.4 Quiz on capacitors and uniform E-field

Question 1. Two parallel plates are separated by 3.0 mm. The voltage between the plates is 180 V as shown.

a. What is the electric field, E, between the plates? Also, draw the E-lines of force due to the plates, with arrowheads showing the direction of the E-field:

E =

0 V   180 V

b. A proton **p**, whose charge is 1.6×10^{-19} C, is in between the plates. What is the force on the proton?

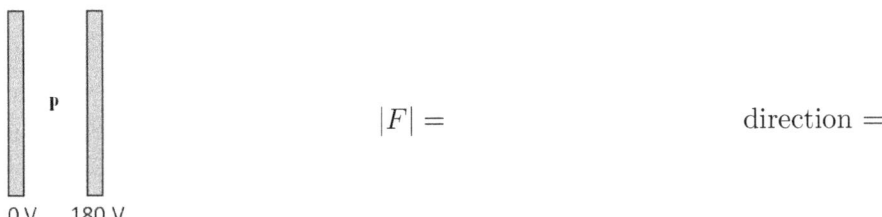

$|F| = $ direction =

0 V   180 V

c. The proton **p** is re-located as shown in the figure below. Now the force —F— on the proton, compared with part b), is   the same   about half   very small   (circle one and explain why)

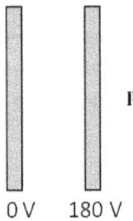

0 V   180 V

d. If the area of the plates is 10 m², what is the capacitance of this capacitor?

6.4. QUIZ ON CAPACITORS AND UNIFORM E-FIELD

6.4.1 Answers to capacitor & E-field quiz

a. $|E| = \Delta V/d = 180$ V $/ 3 \times 10^{-3}$ m $= 60{,}000$ V/m

The E-field is depicted in figure 6.4 a) or b).

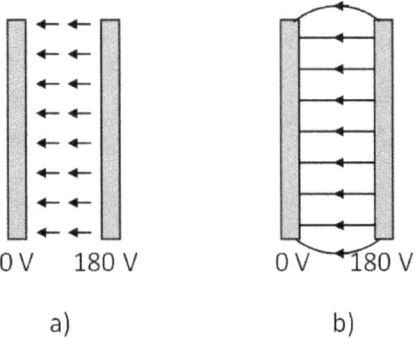

Figure 6.4: Capacitor E-field depicted as a) E-field vectors or b) E-field lines of force

Either answer is acceptable. The bulge in the lines at the ends of the capacitor is called the *fringing field*. This is a fine point; the main idea is that the E-field is essentially uniform between the plates and points in the direction of decreasing voltage.

b. $|F| = |E|\, q = 6 \times 10^4$ V/m $* 1.6 \times 10^{-19}$ C $= 9.6 \times 10^{-15}$ N

c. Much less, because the E-field is almost zero outside a capacitor.

d. $C = A/(4\pi k d) = 10$ m$^2/(4 \cdot 3.14 \cdot 9 \times 10^9 \cdot 3 \times 10^{-3}$ m$) = 29.5$ nF.

6.5 Breakdown Field

When the electric field exceeds a certain value, a single electron or ion can accelerate rapidly enough to initiate an avalanche of current. This 'dielectric breakdown' will produce a spark in a gas, and can destroy nearby material in a solid dielectric. We denote the breakdown field by E_{BD}. For dry air, $E_{BD} = 3 \times 10^6$ V/m. For useful dielectric insulators, the breakdown field can be much more: for mylar, E_{BD} is between 1 and 5×10^8 V/m!

1. A parallel-plate capacitor has an adjustable gap. It is connected to a van der Graaf generator producing a voltage of 70,000 V. What is the minimum gap possible without breakdown in dry air?

2. A parallel-plate capacitor has an air gap of 0.127 mm, and a plate area A = 0.25 m². a. What is the maximum voltage that can be applied before breakdown occurs?

b. What is the maximum amount of charge that can be stored before breakdown occurs?

c. Now suppose the gap in part a) is filled with mylar dielectric. Assume that for mylar $\kappa = 3.1$, and $E_{BD} = 1.2 \times 10^8$ V/m. What maximum voltage, and what maximum charge can be stored using mylar dielectric?

Answers: 1. 0.0233 m   2a. 381 V   2b. 6.63 μC   2c. 1.52 $\times 10^4$ V   822 μC

6.6 Capacitors in series and parallel

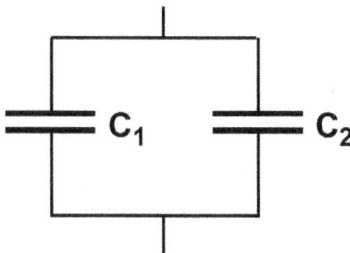

Figure 6.5: Parallel capacitors

When multiple capacitors are arranged in parallel, the same identical voltage must appear across each capacitor. Each capacitor contributes charge to the total charge. When C_1 and C_2 are in parallel, the total load capacitance is

$$C_T = C_1 + C_2 \tag{6.6}$$

When capacitors are lined up in series, the same charge must appear on each capacitor. The total capacitance is less than any one of the individual capacitors.

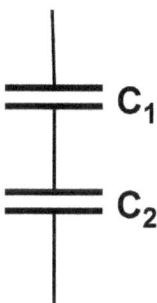

Figure 6.6: Series capacitors

When two capacitors, C_1 and C_2 are in series, the total capacitance is given by

$$\frac{1}{C_T} = \frac{1}{C_1} + \frac{1}{C_2} \tag{6.7}$$

1. We have an 8 μF and a 24 μF capacitor. What is the net capacitance when these are connected in parallel? in series?

Answers: 1. $C_{parallel} = 32~\mu F$ $C_{series} = 6~\mu F$

2. Suppose C_1 is 12 μF and C_2 is 24 μF. a. What is C_T?

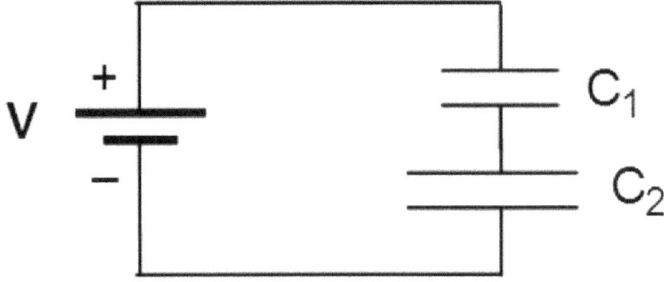

b. If battery voltage V = 9.0 V, what is the charge on C_T?

c. What is the charge on C_1? on C_2?

d. What is the voltage difference across C_1? across C_2?

Answers: 2. a. 8.0 μF   b. 72 μC   c. $Q_1 = Q_2 = 72 \mu$C   d. $V_1 = 6$V, $V_2 = 3$V

6.7 Capacitor energy

If a capacitor C is charged to a voltage V, the energy stored is

$$U = \frac{1}{2}CV^2 \qquad (6.8)$$

where U is the energy stored in the capacitor.

1. A 220,000 μF capacitor is charge with a 12 Volt battery. How much energy is stored?

2. A capacitor has 2 parallel plates separated by a mylar dielectric which is 0.0127 cm thick. The area of the plates is 0.2 m^2. We apply 900 V to the capacitor. How much energy does this store in the capacitor?

3. In the following figure, $C_1 = 20\mu F$ and $C_2 = 40\mu F$. What is the energy stored on each capacitor, and what is the total energy?

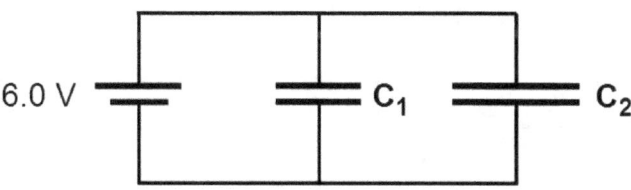

Answers: 1. 15.8 J   2. 0.0175 J   3. $U_{C_1} = 3.6 \times 10^{-4}$ J, $U_{C_2} = 7.2 \times 10^{-4}$ J, $U_T = 10.8 \times 10^{-4}$ J

6.8 Capacitor circuits

1. What is the net capacitance in the following series-parallel capacitor circuit?

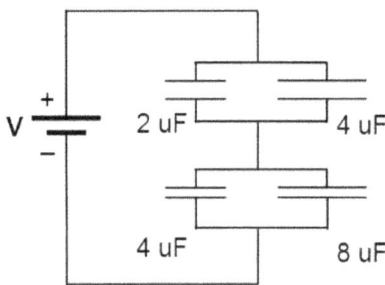

2. a. Find the total capacitance of the following circuit.

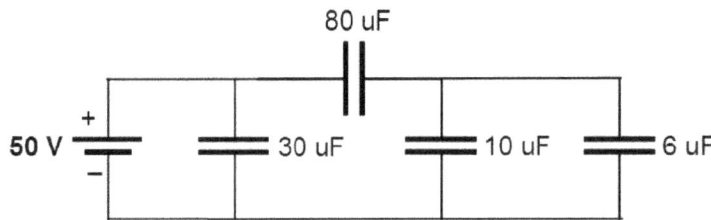

b. Find the total charge stored in the above circuit.

c. Find the total **energy** stored in the above circuit.

Answers: 1. 4.0 μF   2. a. 43.3 μF   b. 2.17 × 10⁻³ C   c. .0541 J

3. a. What is C_T in the following series-parallel capacitor circuit?

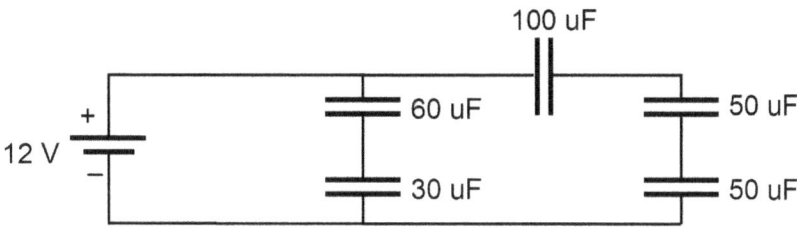

b. What is the voltage across the 100 μF capacitor?

c. What is the energy stored in the entire circuit?

d. What is the energy stored in the 30 μF and the 60 μF capacitors?

Answers: 3. a. 40 μF b. 2.4 V c. 2.88 × 10⁻³ J d. U_{30} = 9.6 × 10⁻⁴ J, U_{60} = 4.8 × 10⁻⁴ J

Chapter 7

RC Circuits

When a resistor is connected across a charged capacitor, the charge leaks through the resistor. The current decays exponentially to zero with a time constant, τ, defined as follows:

$$I = I_0 \, e^{-t/\tau} \quad \text{where} \tag{7.1}$$
$$\tau = R * C \tag{7.2}$$

Therefore when $t = \tau$, current I will have fallen to 37 % its initial value I_0.

1. In the following circuit, switch **S** first connects the battery to the capacitor to charge it. Then it connects the resistor to the capacitor to discharge it.

a. How much time does it take for the voltage on the capacitor to fall to 37% of 12 V = 4.44 V?

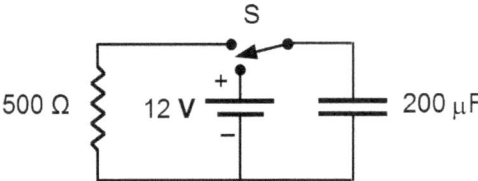

b. How much time does it take for the voltage on the capacitor to fall to 37% of 4.44 V = 1.64 V?

Answers: 1. a. 0.10 s   b. 0.20 s

2. In the below figure, the switch S is initially closed, the capacitors are fully charged, and V = 10 Volts. When S is opened, charge will leak through the resistors. What is the time constant for V to reach 3.7 Volts?

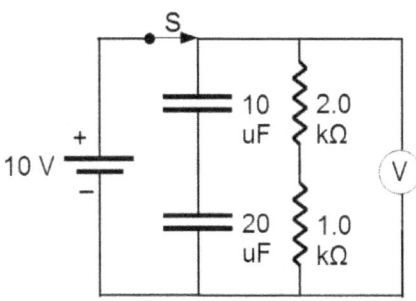

3. When S is opened in the figure below, charge stored on both capacitors will leak through both resistors. What is the time constant for the decay of voltage V?

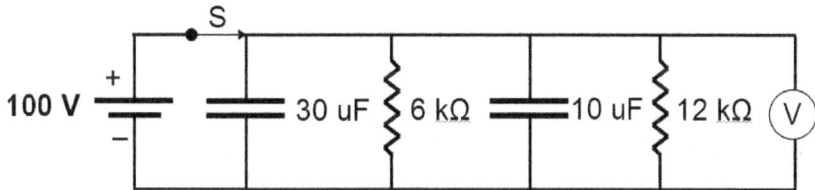

Answers: 3. 0.02 s   4. 0.16 s

Chapter 8

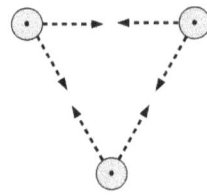

Parallel wires carrying currents out of the page are attracted to each other in pairs.

Magnetic Force Between Currents

If two currents are flowing in two parallel wires, the force between them depends on the values of the two currents, I_1 and I_2, the distance R between the wires, and the length L of the wires:

$$F = \frac{\mu_0}{2\pi}\frac{I_1 I_2}{R} \cdot L \quad \text{where } \frac{\mu_0}{2\pi} = 2 \times 10^{-7}. \text{ Dividing by L}$$

$$f = \frac{\mu_0}{2\pi}\frac{I_1 I_2}{R} \quad \text{where } f = F/L \tag{8.1}$$

is the force per unit length between the wires.

1. Two wires carry a current of 1.0 ampere in the upward direction. They are separated by 1.0 meter and run parallel for a distance of 20 m. What is the magnitude and direction of the force between them?

2. A battery is connected to a long, rectangular loop of wire in the configuration shown. The battery voltage is 1.5 V and the resistance is 0.05 ohm. What is the magnetic force on the wire?

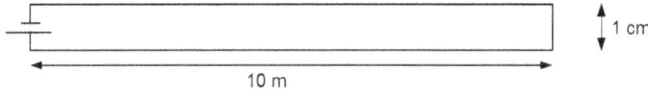

Figure 8.1: Upper and lower wires in the loop carry currents in opposite directions

Answers: 1. 4.0×10^{-6} N horizontal, attractive   2. 1.8×10^{-1} N repulsive

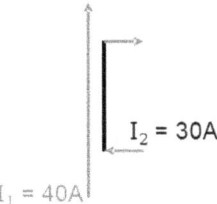

Figure 8.2: Parts of the wire which are not parallel do not contribute to the force

3. In the above figure, wire 2 is 0.8 m long and carries a current $I_2 = 30$ A in the upward direction. It is parallel to, and 1.0 cm away from wire 1.

a. What is the magnitude and direction of the force on wire 2?

b. What is the magnitude and direction of the force on wire 1?

4. 1. Two lightning bolts traveling in the same direction are separated by 0.4 m. One lightning bolt carries a current of 1.5×10^4 A. The other lightning bolt carries a current of 6.0×10^4 A.

What is the force and direction between a 10-meter length of the two lightning bolts?

Answers: 3a. 1.92×10^{-2} N left   3b. 1.92×10^{-2} N right   4. 4.5×10^3 N attractive

8.0.1 More wires

1. Wires a, b, and c are 20.0 m long and carry equal currents of 100 A. What is the net force on each of the wires (Give magnitude and direction)?

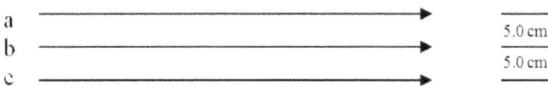

$F_a =$

$F_b =$

$F_c =$

2. The current in wire c is reversed. What is the net force on each of the wires now (Give magnitude and direction)? Use same separation between wires.

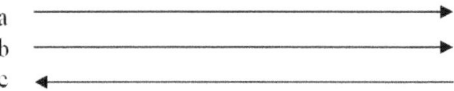

$F_a =$

$F_b =$

$F_c =$

Answers: 1. a. 1.2 N down b. 0 N c. 1.2 N up   2. a. 0.4 N down b. 1.6 N up c. 1.2 N down

8.1 Force of a current on a charge moving parallel

A charge moving parallel or anti-parallel to a current can itself be viewed as a current. Then there exists a force between the two currents. If the charge is q, moving at speed v, q*v has the units, Coulomb*meter/sec = Amp-m. The force equation (8.1) becomes

$$F = \frac{\mu_0}{2\pi} \frac{Iqv}{R} \qquad (8.2)$$

8.1.1 positive charge

If the charge is positive and the velocity is in the same direction as the current, the force is attractive. If the positive charge moves antiparallel to the current, the force is repulsive.

8.1.2 negative charge

If the charge is negative, it behaves as a current opposite in direction to the velocity. The force between a current and a negative charge moving parallel to the current is therefore repulsive. A negative charge moving opposite to teh direction of a current is attracted to the current.

1. A vertical wire carries an upward current of 20.0 A. An electron is located 1.0 cm to the right of the wire, and is moving vertically. The charge on an electron is -1.6×10^{-19} C. For each of the three velocities given what is the force on the electron (Give magnitude and direction)?

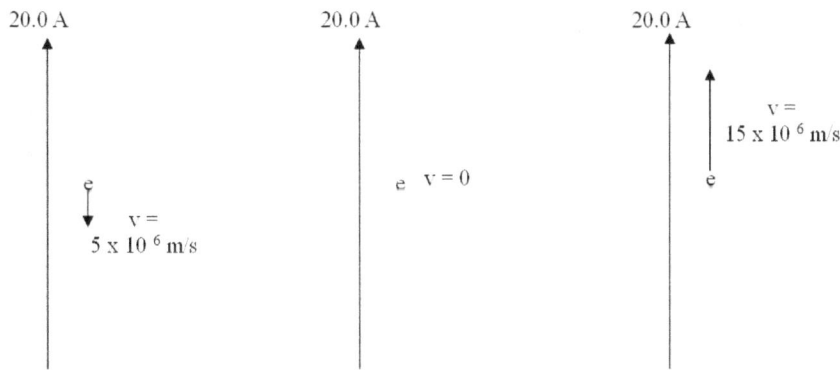

Answers: 1. a. 3.2×10^{-16} N attractive   b. 0 N   c. 9.6×10^{-16} N repulsive

2. In each case of the charge and wire in problem (1) above, what would be the effect on the force a. if the charge were moving twice as fast?

b. if the charge were twice as far from the wire?

c. if the current in the wire were twice as large?

3. Two wires carry 40 A currents in opposite directions. The wires are 1.0 cm apart. What will be the effect on an electron moving at a speed of 3×10^5 m/s, if it is moving parallel to the wires, but half way in between the two? Show the path of the electron and calculate its acceleration. ($q_e = 1.6 \times 10^{19}$ C, and $m_e = 9.11 \times 10^{-31}$ kg).

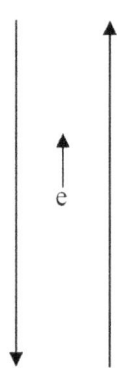

8.2 B-field around a wire

Just as the electric field E mediates the force between charges, the magnetic force field B mediates the force between currents. A straight wire carrying a current I creates a magnetic field B, given by

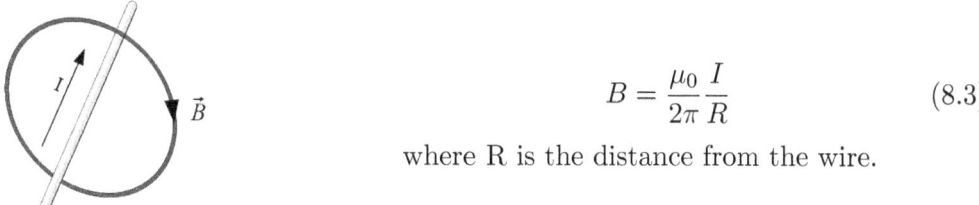

$$B = \frac{\mu_0}{2\pi} \frac{I}{R} \qquad (8.3)$$

where R is the distance from the wire.

B-field direction Right hand thumb rule: Place the right hand thumb along the direction of positive current flow. The B-field points along concentric circles that are centered on the wire. B-field points in the same direction as the fingers are pointing around the wire.

1. A wire carries a current of 40 A in the upward direction. Draw the magnetic field and find its value and direction at point **a)** 1 cm from the wire; and at point **b)** 3 cm from the wire?

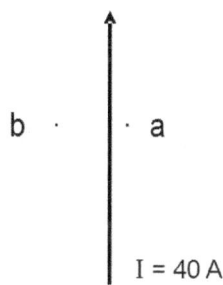

2. A current I points into the page as shown. Draw arrows to indicate the direction and magnitude of the B-field at points A, C, and D. (Hint: Each arrow must be tangent to a circle centered on the wire.)

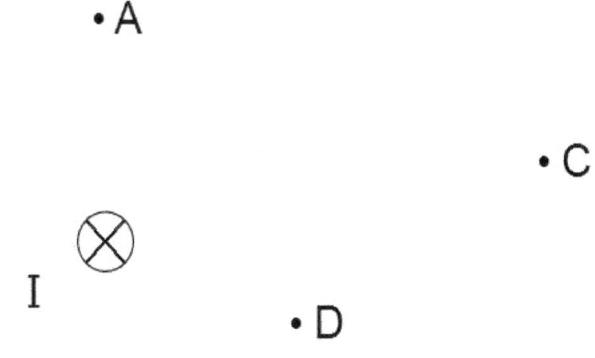

Answers: 1. a. 8×10^{-4} T into the page b. 2.67×10^{-4} T out of the page

3. In the drawing below, wire1 carries a current of 40 A in the upward direction. a. Consider first the magnetic field due to wire1. Find its value and direction at point **a)** 1 cm from the wire.

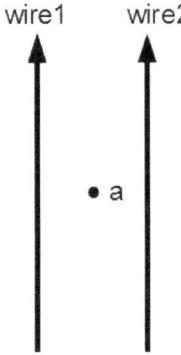

b. Wire2 also carries a current of 40 A in the upward direction. Consider the magnetic field due to wire2 and find its value and direction at point **a)** assuming it is also 1 cm from wire2.

c. What is the net B-field at **a)** due to both wire1 and wire2?

d. How do parts b) and c) above change if wire1 has a downward directed current instead of upward?

Answers: 3. a. 8×10^{-4} T into the page b. 8×10^{-4} T out of the page c. 0 T d. $B_{net} = 16 \times 10^{-4}$ T out of page

8.2. B-FIELD AROUND A WIRE

4. Two wires are carrying identical currents in the configuration shown below. They are parallel and a distance **x** meters apart at the right side of the figure. The wires separate and the lower wire continues by itself on the left side of the figure. Point **b** is halfway between the two currents.

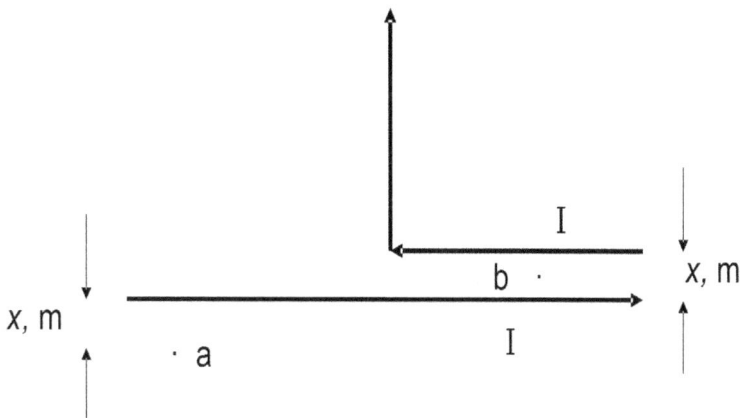

Question: If the B-field at **a** is 0.01 T, what is the B-field at **b**? What are the directions of the two B-fields?

5. Two long wires carry currents of 40 A and cross at right angles as shown. Point **A** is 1.0 cm from the horizontal wire, and 2.0 cm from the vertical wire. What is the magnetic field at point **A**? Give magnitude and direction.

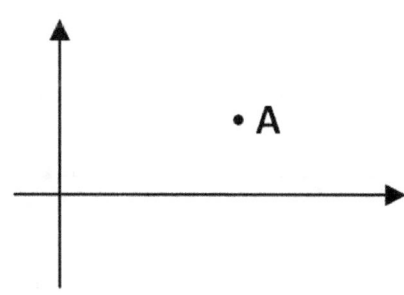

Answers: 4. 0.04 T   b out of page   a into page   5. 4×10^{-4} T   out of page

6. Three wires cross at 60° angles but do not short to each other. Each wire carries 90 A. Point **P** is a distance 0.45 cm from each of the three wires.

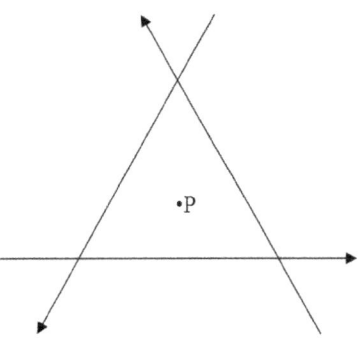

a. What is the net B-field at point **P**?

b. The right-side current is reversed in direction. What is the net B-field at **P** now?

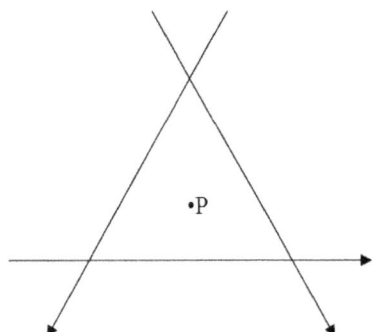

Answers: 6. a. 0.012 T, out of page   b. 0.004 T, out of page

8.2. B-FIELD AROUND A WIRE

7. Three long wires carry currents of 40 A and are arranged in an equilateral triangle as shown. Point **A** lies in an exterior angle and is 3.0 cm from each of two of the wires, and 5.0 cm from the more distant wire. What is the magnetic field at point **A**? Give magnitude and direction.

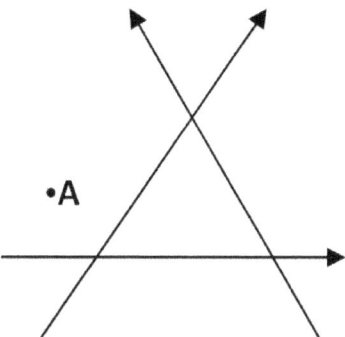

8. Four long wires carry currents of 20 A and are arranged in a square as shown. The square has a side s = 1.5 cm. What is the magnetic field at the center of the square, point **A**? Give magnitude and direction.

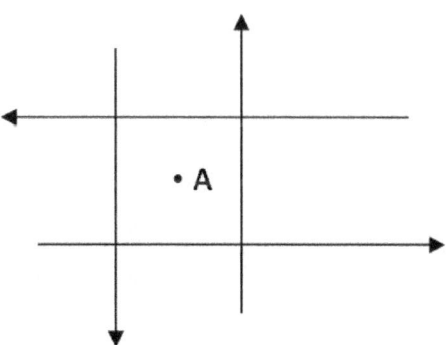

Answers: 7. 6.93×10^{-4} T out of page 8. 2.13×10^{-3} T out of page

Chapter 8. Magnetic Force Between Currents

9. The two wires cross at an 45° angle but do not short to each other. Each wire carries 60 A. Points **P** and **Q** are each distance 0.8 cm from both wires.

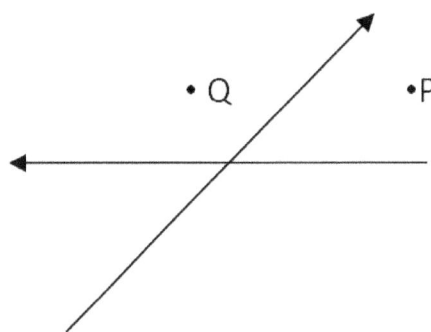

a. What is the net B-field at point **P**?

b. What is the net B field at point **Q**?

Answers: 9. a. 3.0×10^{-3} T into page b. 0 T

8.2.1 More B-field around parallel wires

1. Two wires are 2.0 cm apart, carry parallel currents of 50 A each, and both are out of the page.

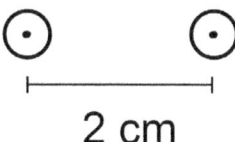

a. What is the magnetic field exactly half way in between the two wires?

b. What is the magnetic field 5.0 cm to the left of the lefthand wire?

c. What is the magnetic field 5 cm to the right of the righthand wire

d. Draw arrows at the points defined by a), b) and c) above to show the direction and relative magnitude of the fields you have calculated.

Answers: 1. a. 0 T   b. 3.43×10^{-4} T down   c. 3.43×10^{-4} T up

2. Referring to the previous page, now reverse the current in the lefthand wire.

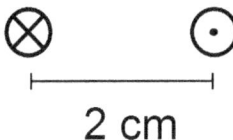

a. What is the magnetic field exactly half way in between the two wires?

b. What is the magnetic field 5 cm to the left of the lefthand wire?

c. What is the magnetic field 5 cm to the right of the righthand wire?

d. Draw arrows at the points defined by a), b) and c) above to show the direction and relative magnitude of the fields you have calculated.

Answers: 1. a. 2.0×10^{-3} T down   b. 5.71×10^{-5} T up   c. 5.71×10^{-5} T down

8.3 Quiz on force between wires & B-field around a wire

1. A and B are two wires separated by 2.0 cm and carrying parallel currents. They carry 4.0 A and 10.0 A, resp. The wires are 52 m in length. What is the magnitude and direction of the force on wire B?

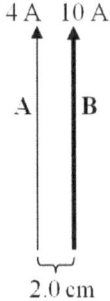

2. B and C are separated by 2 cm and carrying **opposing** currents. They carry 10.0 A and 6.0 A, resp. The wires are 52 m in length. What is the magnitude and and direction of the force on wire B?

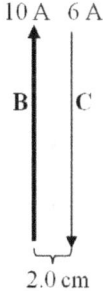

3. B and C are two very long wires separated by 2 cm and carrying **opposing** currents. They carry 10.0 A and 6.0 A, resp. Point **p** is half way between the wires. What is the magnitude and direction of the B-field at point p?

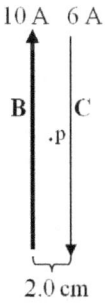

4. A third wire parallel to B and C, carrying current in the upward direction, is placed somewhere in between the two wires. What is the direction of the force on this third wire?

8.4 Answers to quiz on force between wires & B-field around a wire

1. The currents are parallel, so direction of the force is *attractive*. It is also correct to say the left wire has a rightward force and the right wire has a leftward force. Magnitude is $F = (\mu_0/2\pi)I_1 I_2 L/r = 2 \times 10^{-7} \cdot 4 \text{ A} \cdot 10 \text{ A} \cdot 52 \text{ m}/0.02 \text{ m} = 0.0208 \text{ N}$.

2. The currents are anti-parallel, so the force is repulsive. The product of the currents is 1.5× the product of currents in question 1) above, so the force is 1.5× greater, or 0.0312 N.

3. B-field at point **p** due to wire **B** is $B = (\mu_0/2\pi)I/r = 2 \times 10^{-7} \cdot 10 \text{ A}/0.01 \text{ m} = 2.0 \times 10^{-4}$ T. Direction is *into* the page. B-field at point **p** due to wire **C** is 0.6× as large or 1.2×10^{-4} T. Both fields, from B and from C, are directed into the page, so they add to give $B = 3.2 \times 10^{-4}$ T.

4. Left. The third wire is pushed in the same direction by both **B** and **C**.

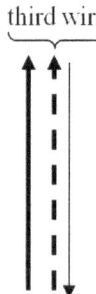

Figure 8.3: Draw all force vectors and show that both B and C push the 3rd wire to the left.

8.5 B-field around a current loop

A straight, current-carrying wire bends into a curved wire, and finally into a complete loop. The figure below shows how the B-field depends on the shape of the wire.

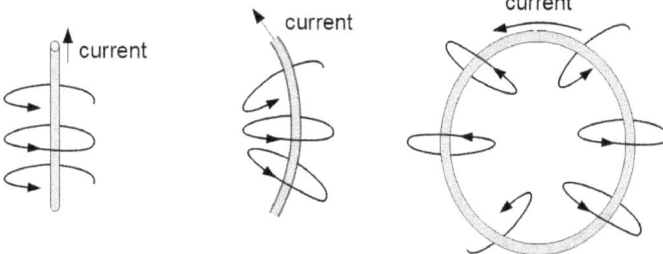

The B-field first of all concentrates on the inside of a curve. Then, as the loop is closed, the Right Hand Thumb Rule shows that all the field is entirely in one direction coming out of the inside of the loop. (This is similar to the air currents in a smoke ring – scientifically speaking, a 'vortex ring'.)

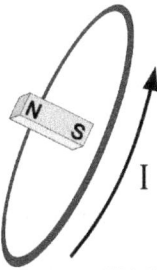

Figure 8.4: Current loop imagined as a properly oriented bar magnet

The North end of a magnet also emits B-field. The lines curve around and return to the South end of the magnet. Therefore we can depict the current loop as being similar to a bar magnet. This is illustrated in Figure 8.4.

1. A magnet is brought near a current-carrying loop as shown. What will be the direction of the force on the loop, and its subsequent motion?

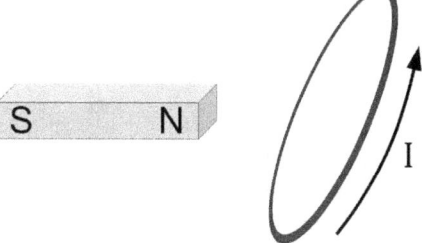

Chapter 9

B-field Force on a Current

A current **I** immersed in a B-field **B** experiences a force. The force is maximum if B and I are perpendicular, and has the value

$$\begin{aligned} F &= BIL \quad \text{L is the length of wire. Dividing by L,} \\ f &= BI \quad \text{where } f = F/L \text{ is the force per unit length} \end{aligned} \quad (9.1)$$

Force direction This is given by the Right Hand Rule:

1. Extend fingers so they point in the direction of I
2. Rotate hand until palm faces in the direction of B, and fingers can bend from I to B
3. Then the thumb shows the direction of F

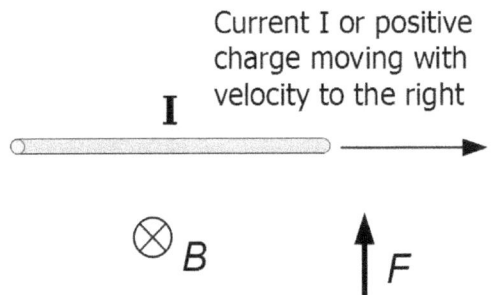

Figure 9.1: Example of force due to B-field interacting with a current-carrying wire

9.1 Magnetic force magnitude

1. An everyday example is the force of the Earth's magnetic field (5×10^{-5} T) acting upon a power line. The power line is 200 m long and carries a current of 5000 A. Assume the power line runs East-West, so that the Earth's B-field is perpendicular to the wire. Positive current flow is toward the East. What is the force on the wire?

Answer: $F = BIL = 5 \times 10^{-5}$ T $*$ 5000 A $*$ 200 m $= 50$ N.

I not $\perp$ to B: What if the current is not perpendicular to the B-field? Then Equation 9.1 becomes

$$F = BIL \sin \theta \tag{9.2}$$

where θ is the angle between I and B.

2. A current of 10 ampere is carried by a wire perpendicular to a 1.38 T field (a very strong permanent magnet). The magnet is only 2 cm in diameter, so the part of the wire in the magnetic field is only 2 cm. a. What is the force on the wire?

b. Sketch the configuration and draw an arrow to indicate the direction of the force.

c. A current of 10 amperes is carried by a wire that is *parallel* to a 1.38 T field. What is the force on the wire?

Answers: 2. a. 2.76×10^{-1} N c. 0 N

3. The B-field is 0.15 T and the wire is 3.0 m long and carries a current of 24 A

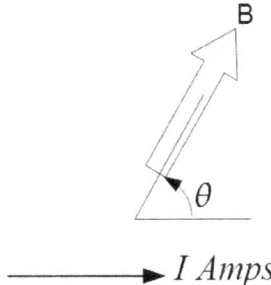

Figure 9.2: Define θ so B is up the page

a. If $\theta = 0°$ what is the force on the current?

b. If $\theta = 30°$ what is the force on the current?

c. If $\theta = 150°$ what is the force on the current?

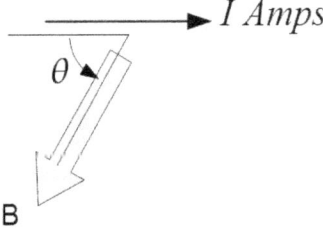

Figure 9.3: Re-define θ: B is now *down* the page

d. If $\theta = 0°$ what is the force on the current?

e. If $\theta = 30°$ what is the force on the current?

f. If $\theta = 150°$ what is the force on the current?

Answers: 3. a. 0 N   b. 5.4 N   c. 5.4 N

9.1. MAGNETIC FORCE MAGNITUDE

9.1.1 More magnetic direction

1. Find the direction of the Force on the wire with the current I.

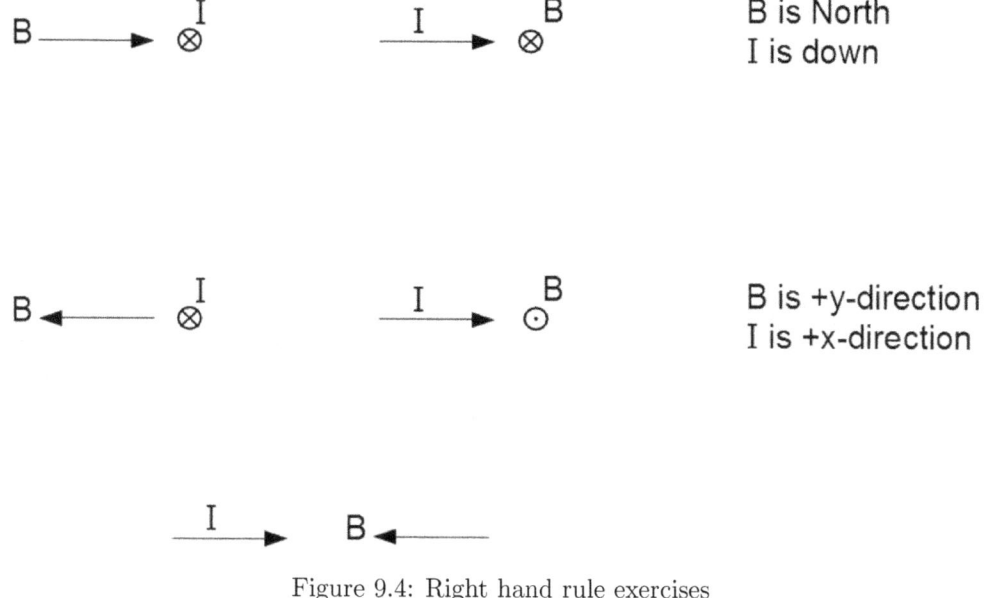

Figure 9.4: Right hand rule exercises

2. Find the direction of the magnetic field given current I and force F. Assume B is at right angles to I.

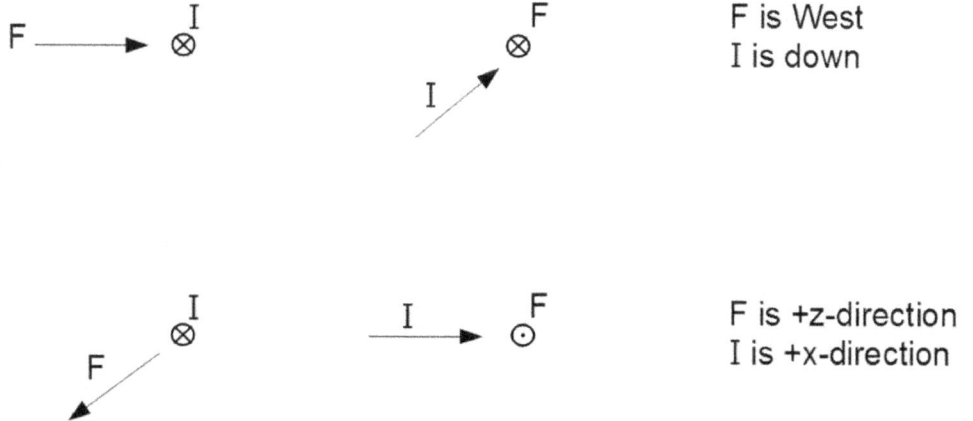

3. Draw B assuming B is at 45° angle to I:

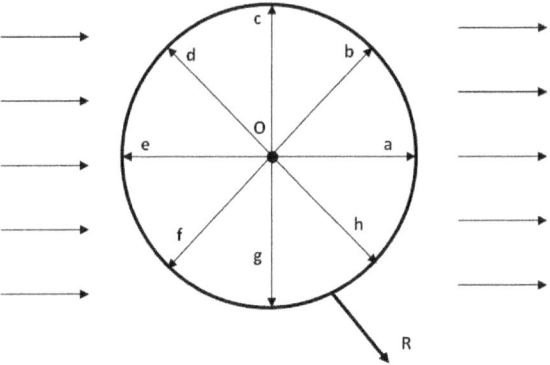

Figure 9.5: Equal currents going out the spokes to the rim

4. A current of 80 A enters a wheel at the hub, **O**, divides into the spokes a-h as shown, rejoins on the rim and leaves the rim through the connection at R. The radius of the wheel is 25 cm. The spokes are separated by 45° angles. A magnetic field having magnitude 0.14 T is pointed to the right, and is in the plane of the wheel. Find the force on each spoke

|F| direction

a)

b)

c)

d)

e)

f)

g)

h)

5. What is the net force (**magnitude** and **direction**) on all the spokes added together?

Answers: 4. a. 0   b. 2.47×10^{-1} N   c. 3.5×10^{-1} N   5. 0 N

9.2 Net force on a current loop

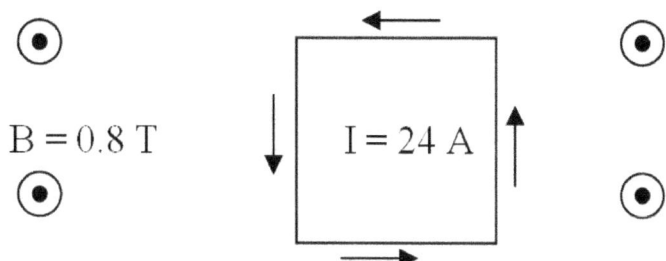

1. A square loop is 0.20 m on a side and carries a current I = 24 A in a counterclockwise direction. The coil is immersed in a uniform field of 0.8 T, directed out of the page.

a. What is the value and direction of the force on each side of the coil?

top

bottom

right

left

b. What is the net force on the loop?

2. Now the same loop is immersed in a **horizontal** uniform field of 0.8 T.

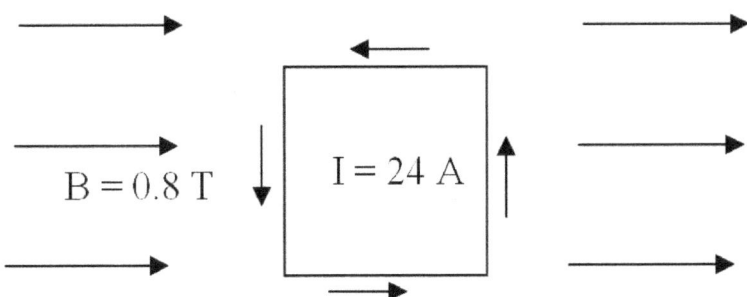

a. What is the value and direction of the force on each side of the loop?

top

bottom

right

left

b. What is the net force on the coil?

9.3 B-force on a moving charged particle

A charged particle moving in a uniform B-field experiences a sideways force. In analogy to Equation (9.1), the force is given by

$$F = qvB \qquad \text{force on a charged particle in a B-field} \qquad (9.3)$$

The direction of the force is given by the Right Hand Rule – in which the velocity vector of a positive charge takes the place of the wire direction in Figure 9.1. Negatively charged particles experience a force in the opposite direction – i.e., the direction of current is taken to be exactly *opposite* the velocity of a negative charge. The particle moves in a circular path.

Assume the particle has mass m, and charge q. The force must equal the mass times the centripetal acceleration, so

$$qvB = \frac{mv^2}{R} \qquad \text{and, solving for R}$$
$$R = \frac{mv}{qB} \qquad (9.4)$$

1. In a mass spectrometer, a $^4\text{He}^+$ ion (mass $4 \times 1.67 \times 10^{-27}$ kg) is moving at a speed of 1×10^5 m/s. The charge is $+1 \times 1.6 \times 10^{-19}$ C.

a. What is the radius of its orbit in a magnetic field of 0.1 T?

b. What is the orbit size for a doubly ionized $^4\text{He}^{++}$?

Answers: 1. a. 4.18×10^{-2} m   b. 2.09×10^{-2} m

2. The solar wind contains large amounts of protons emitted from our Sun by solar flares. The protons move at a velocity $v = 0.60 \times 10^6$ m/s. The mass of the proton is 1.67×10^{-27} kg. The proton encounters the magnetic field of the Earth, $B = 5.5 \times 10^{-5}$ T.

The Earth's magnetic field protects us by deflecting the protons away from us into circular, spiraling paths.

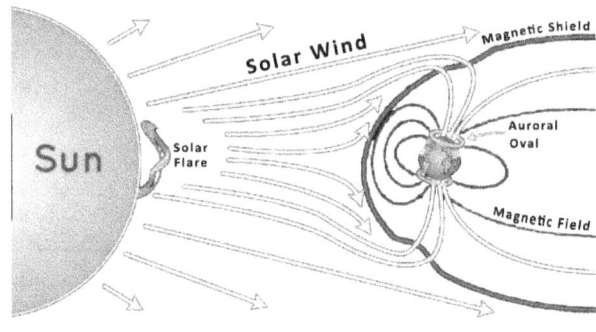

a. Given the speed and mass of these particles, what is the radius of their circular path?

b. When solar wind particles hit the upper atmosphere, they create a shower of excited atoms and ions that re-emit the energy as light. What is the name of this phenomenon, and where can it be found?

3. Electron beams have been used in cathode ray display tubes (CRTs) and x-ray tubes. Proton beams are used in high energy particle accelerators (e.g., synchrotrons).

A magnet can be used to steer the beam. An electron beam in a CRT moves toward the screen with a velocity v = 4.2×10^7 m/s. A bending magnetic field is applied perpendicular to the beam. It causes the beam to deflect with a radius of curvature 0.75 m. What is the magnitude of B-field that would cause this radius of curvature? [Use the electron mass $m_e = 9.1 \times 10^{-31}$ kg.]

Answers: 2. a. 1.14×10^2 m   3. 3.19×10^{-4} T

9.4 B-force on a current loop

Imagining a current loop as a little magnet is a quick way to analyze forces and torques on a current loop. Another valuable method is to solve such problems using $F = BIL$ and the Right Hand Rule. Use both methods in the following problem to confirm they give similar answers.

In each of the following figures, a-e, what will be the motion of current loop T? Assume any magnets or loops other than T are held fixed in position.

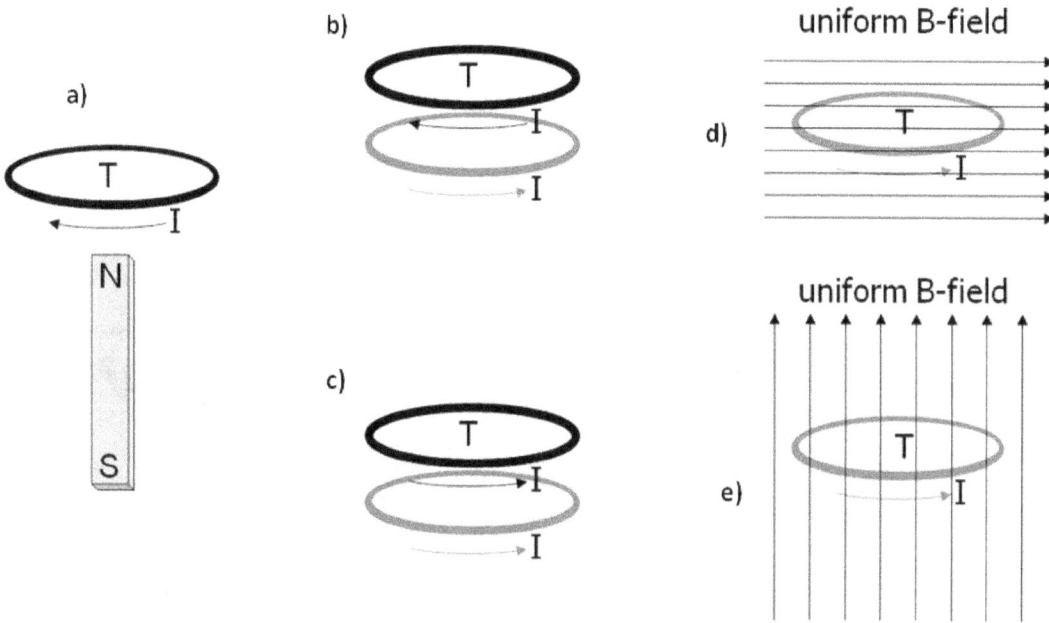

9.5 Torque on a current loop

A uniform field exerts no net force on a current loop. It does, however, exert a torque on a current loop. In the following equation, N is the number of turns in the coil, A is the area of the coil, and B is the magnetic field. Torque is given by

$$\tau = NIAB \quad \text{maximum torque}$$
$$\tau = NIAB \cos\Theta \quad \text{torque vs. angle} \tag{9.5}$$

Angle θ is the angle between the B-field direction and the *plane* of the coil.

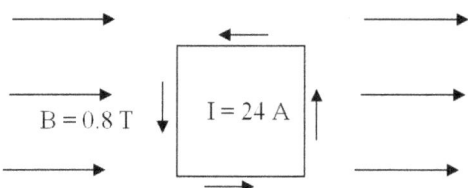

1. A square loop is 0.2 m on a side and carries a current I = 24 A in a counterclockwise direction. It is immersed in a horizontal uniform field of 0.8 T.

a. What is the value and direction of the force on each side of the loop?

top

bottom

right

left

b. Use part a) above to find the net torque on the loop. (Find torques about a vertical axis through the center of the loop.)

$\tau =$

c. Use the formula, $\tau = NIAB \cos\theta$ to find the net torque on the loop:

$\theta =$ $\qquad$ $\tau =$

Answers: 1. c. $\theta = 0°$ $\quad \tau = 7.68 \times 10^{-1}$ N-m

9.5. TORQUE ON A CURRENT LOOP

d. What is the torque on the loop with B turned as in the figure below?

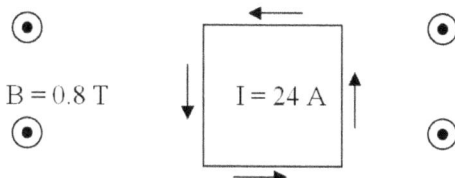

2. The Earth's magnetic field is 5.5×10^{-5} T. Assume we are at the equator and there is zero dip angle. A square wire coil that is 8.0 m long and 6.0 m wide is lying flat on the ground. The coil has 25 turns and is energized with a current of 10 A in the wire.

a. What is the torque on the coil due to the Earth's field? Do this in two ways: (1) Find the force on each of the sides of the rectangle; pick a horizontal axis of rotation; and add up the torque due to each force. (2) use equation (9.5).

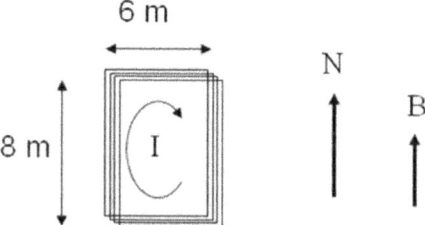

b. What is the net force on the coil?

3. Turn the coil 90° and repeat 2a) and 2b) above.

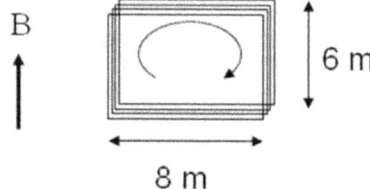

Answers: 1. d. $\theta = 0°$ 2. a. $\tau = 6.6 \times 10^{-1}$ N-m   3. a. 6.6×10^{-1} N-m

9.6 Quiz on B-field force and torque

1. The figure shows the B-field lying in the plane of a loop in the shape of a right triangle. The loop has 55 turns. In each turn, current I = 35 A, and B = 0.22 T. Side Q = 30 cm, S = 40 cm, and side R = 50 cm.

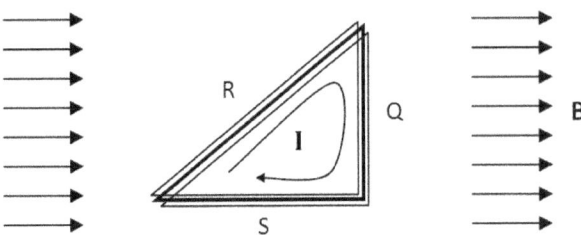

a. Fill in the table with the magnitude and direction of the force on each side of the loop:

side	magnitude	direction
Q		
S		
R		

b. (1 point) What is the magnitude of the net torque on the loop?

c. (extra credit) What is the magnitude of the net force on the loop?

Answers: a. Q: 127 N out of page   R: 127 N into page   S: 0 N   b. 25.4 N-m   c. 0

9.6.1 Answers to Quiz on B-field force and torque

a. Side Q: Use the formula, $F = BIL \sin \theta$, and note that the current is really the sum of currents in all the 55 turns of the coil. Therefore, $F = 0.22$ T $* 55 * 35$ A $* 0.3$ m $* \sin 90° = 127$ N.

Side R: $F = 0.22$ T $* 55 * 35$ A $* 0.5$ m $* 0.6 = 127$ N, because $\sin \theta = 3/5$.

Side S: F = 0, because B is parallel to the current.

The table appears as follows:

side	magnitude	direction
Q	127 N	out of page
R	127 N	into page
S	0 N	

b. For torque, use $\tau = NBIA \cos \theta$, where θ is now defined as the angle between the B-field and the plane of the coil – i.e., $\theta = 0$. Then $\tau = 55 * 0.22$ T $* 35$ A $* ((0.4 * 0.3)/2)$ m$^2 * 1 = 25.4$ N-m.

c. The forces on Q and R are equal in magnitude but opposite in direction. Sum of forces is zero.

Chapter 10

The Solenoid

Section 8.5 showed how the B-field lines densify inside a curved wire, and the B-lines form a vortex ring around a current loop. A long coil is called a **solenoid**. As in the loop, the B-lines are intensified inside the coil. The B-lines spread out and circulate back around the outside of a current loop, and, similarly, they curve around the outside of a solenoid. But the density of B-lines gets much lower outside a long solenoid. For practical purposes B = 0 outside the solenoid. Inside the solenoid, B is given by

$$B = \mu_0 I N_{tpm} \qquad \text{B-field of a solenoid} \qquad (10.1)$$

N_{tpm} is the density of coil turns, measured as the number of turns per meter.

1. A long solenoid has 5000 turns of wire. The solenoid is 4.0 cm long. The current in the wire is 16 A.

a. What is the magnetic field inside the coil?

b. What is the B-field outside the coil?

Answers: 1. a. 2.513 T   b. 0

B-field Direction: If the right hand thumb is aligned with the flow of current through any loop in the solenoid, and the fingers are curled inside the solenoid, then the B-field points in the direction of the fingers.

2. Draw the B-field lines around a coil that is carrying a current as shown. The density of field lines should indicate the relative strength of the B-field, and use arrowheads to show the direction of the B-field.

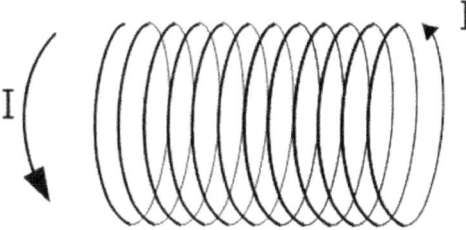

3. A solenoid has 2500 turns of copper wire wound on a cylinder which is 10.0 cm long and 8.0 cm² in cross sectional area. It is carrying a current of 16.0 A. Note that $\mu_0 = 2 \times 2\pi \times 10^{-7}$.

a. What is the B-field in the center of the solenoid?

b. What is the B-field outside of the solenoid, away from the ends?

c. What is the inductance, L, of this solenoid, and how much energy is stored in the inductor? (The units of inductance are henries, H). The formula for inductance is $L = \mu_0 N_{tpm}^2 lA$. Here l is the length of the solenoid, and A is the area of the cross section. The energy $= \frac{1}{2}LI^2$.

L =   Energy =

Answers: 3. a. .503 T   b. 0   c. 6.29×10^{-2} H   8.04 J

10.1 Solenoid B-field force

1. A 'long' solenoid has 10,000 turns. It is 0.12 m in diameter, and 0.40 meter long. The current in the solenoid is 6.0 A. A wire carrying 90 A passes through the center of the solenoid and at right angles to its long axis, as shown.

a. What is the force on the wire and what is its direction? (Hint: First find the field inside the solenoid and then work out the force on the 90 A current.)

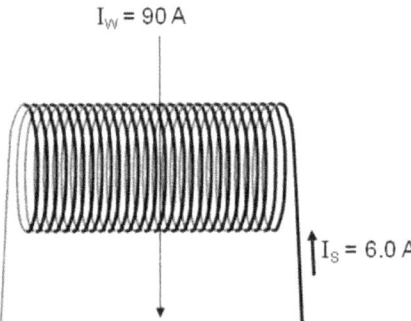

2. A 'long' solenoid has 28,500 turns. It is 0.20 m in diameter, and 1.30 meter long. The current in the solenoid is 22.0 A. A square coil, 0.10 meter on a side, contains 23 turns. It is sitting far outside the solenoid as shown. The square coil has a current of 5.0 A circulating in it.

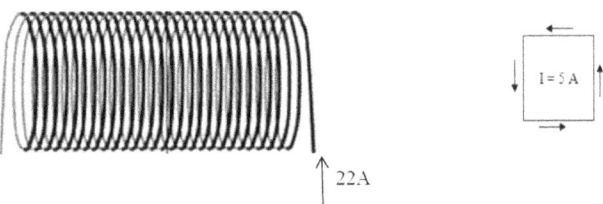

a. What is the torque on the square coil, and what is its direction?

b. Now move the square coil, preserving its current and orientation, into the middle of the solenoid. What is the torque on the square coil now, and what is its direction?

Answers: 1. a. 2.04 N into the page   2. a. 0 N-m   b. 0.697 N-m right side up, left side down

10.2 Solenoid proportional reasoning

1. Problem: A solenoid has 10 amperes going through it. The solenoid has length of 0.15 m. It has 80 turns in its entire length.

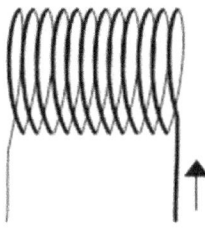

a. What is the magnetic field inside the solenoid?

b. What is the magnetic field outside the solenoid?

c. What is the direction of the magnetic field inside the solenoid?

2. Two solenoids that are identical to the one above are aligned and placed end to end. They are hooked up in series, and the same current, 10.0 A goes through each. What is the magnetic field inside this new, combined solenoid?

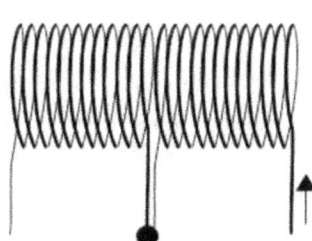

Answers: 1. a. 6.70×10^{-3} T b. 0 c. left 2. 6.70×10^{-3} T

3. Now the original solenoid [previous page, problem #1], carrying the same 10 Amperes, is stretched out to twice its original length. What is the B-field inside the solenoid now?

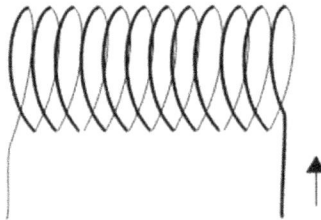

4. Now the original solenoid in problem (1) – 0.15 m in length and carrying the same 10 A – is shrunk in diameter from its original 0.04 m to 0.02 m. What is the B-field inside the solenoid now?

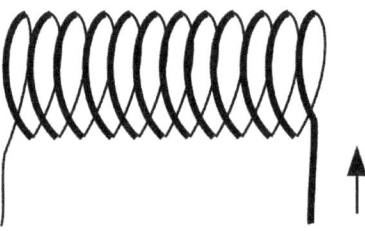

5. A solenoid has an unknown number of turns, N. Its length is 12 cm, and diameter is 3.5 cm. When a current I = 6.9 A passes through the solenoid, it produces a magnetic field B = 0.044 T. If the number of turns is tripled, but the length of the coil remains the same, what is the new B-field?

Answers: 3. 3.35×10^{-3} T   4. 6.70×10^{-3} T   5. 0.132 T

Chapter 11

Faraday's Law and Lenz's Law

A changing magnetic field causes an electromotive force, or emf. Faraday's Law governs the size of the emf, and Lenz's Law provides the direction.

Suppose a wire loop having area **A** is immersed in a uniform B-field, and the B-field is oriented at right angles to the plane of the loop. Suppose the B-field increases between time (a) and time (b). Then an **emf**, or voltage, is generated in the wire, causing a current to flow. The size of the emf is

$$emf = -\frac{\Delta BA}{\Delta t} \qquad \text{Faraday's Law} \qquad (11.1)$$

More generally, the *B-flux* is the number of B-lines of force penetrating the loop at any angle. A flux of 1 Weber (Wb) corresponds to a magnetic field of 1 Tesla penetrating an area of 1 m^2 at right angles, or 1 T-m^2 =1 Wb.

$$emf = -\frac{\Delta \Phi}{\Delta t} \qquad \text{where } \Phi \text{ is the flux} \qquad (11.2)$$

If the B-field penetrates the area of the loop at an angle, then fewer lines of force penetrate the area, and the flux is correspondingly smaller. As an extreme example, if the B-field is parallel to the plane of the loop, then there is zero flux through the loop.

If the loop is replaced by a coil having N turns, then (11.2) becomes

$$emf = -N\frac{\Delta \Phi}{\Delta t} \qquad \text{where N is the number of turns} \qquad (11.3)$$

1. A 20-turn coil has an area 10 cm^2. If the magnetic field increases from 0 to 0.5 Tesla through the coil in 1/60 s, what is the EMF generated in the coil?

Answer: Each turn of the coil develops an emf and these voltages add in series. The result is
EMF = 20 * 0.5 T * 10^{-3} cm^2 * 60 sec^{-1} = 0.60 V.

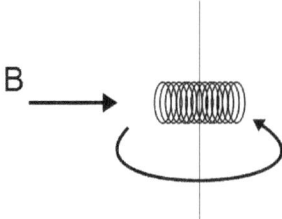

2. Inside a solenoid having 10,000 turns and 10 cm² cross-sectional area, a B-field of 0.05 T is aligned along the axis. The field is held constant in size and fixed in direction. In 0.3 second the solenoid flips around so its long axis points in the exact opposite direction. What is the size of the emf generated in the solenoid?

Hint: The flux Φ is a vector with direction, so the *change* in flux, $|\Delta\Phi| = |\Phi_{final} - \Phi_{init}| = 2 \times |\Phi_{init}|$.

3. A wire bent into a large, circular loop slides across the face of a magnet. At the beginning of the motion, the magnetic field in the loop is zero, and at the end of the motion the loop circles the entire pole face of the magnet. The loop area is 300 cm². The area of the pole faces is 200 cm². The B-field of the magnet is 1.2 T. If the loop takes 0.5 s to complete this motion, what is the average emf induced in the loop?

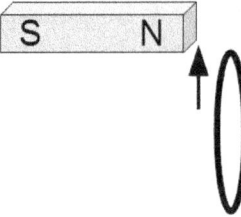

Figure 11.1: Loop quickly moves up to magnet pole.

4. A circular coil has 4 turns of wire and is immersed in a perpendicular, uniform magnetic field, B = 0.21 T. The area of the coil is A = 0.44 m². Suddenly, the coil unravels and reforms into a loop containing a single turn. The new loop diameter is 4× that of the original coil. This change occurs in 0.17 second. What is the average emf produced in the coil as it unravels in the B-field?

Answers: 2. 3.33 V   3. 0.048 V   4. 6.52 V

EMF Direction: **Lenz's Law** determines the direction of the Faraday-induced voltage. It states that when there is a change in B-flux Φ through a loop, the EMF in the loop *opposes* the change in B-flux.

1. A magnet is moved towards a fixed conducting loop, as shown below. The magnetic field due to magnet will increase as the magnet gets closer to the loop.

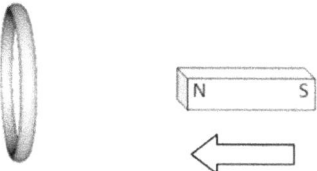

a. Use Lenz's Law to determine the direction of the current induced in the loop. Draw an arrow labeled *I* next to the loop showing the direction of current flow as the magnet approaches.

b. Draw a little magnet corresponding to the magnetic field produced by the current I induced in part a) above. Label the north and south ends of the little magnet with an '**N**' and an '**S**'.

c. What is the direction of the magnetic force experienced by the magnet as it approaches the loop? This force will cause the magnet to   **slow down**   **speed up**   **continue unchanged**   as it approaches the loop (circle one).

2. Now the magnet is moved away from the fixed conducting loop, as shown below. The magnetic field due to magnet will decrease as the magnet gets farther away from the loop.

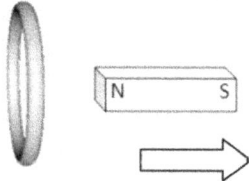

a. (1) Use Lenz's Law to find the direction of the current induced in the loop, (2) draw a little magnet corresponding to the magnetic field produced by the induced current, and (3) label the north and south ends of the little magnet with an '**N**' and an '**S**'.

b. What is the direction of the magnetic force experienced by the magnet as it pulls away from the loop? This force will cause the magnet to   **slow down**   **speed up**   **continue unchanged**   as it moves away from the loop (circle one).

3. A fixed, strong magnet near a conducting metal plate can act as a magnetic brake to slow down motion of the metal plate. Explain:

CHAPTER 11. FARADAY'S LAW AND LENZ'S LAW

11.1 Faraday Applications

1. Principle of the Magneto The magnetic field at the North and South poles of a bar magnet has a magnitude of 0.23 T. The area of each magnet pole is 2.22 cm^2. A coil of wire containing 5000 turns is located at the North end of the magnet. The magnet starts out pointing along the axis of the coil, with the North end of the magnet right against the plane of the coil.

Suddenly, the magnet rotates 180°, so that the South end of the magnet is against the coil. This motion takes place in 0.015 s.

a. What is the average voltage induced in the coil during this time interval?

b. Is the area of the coil important, and why?

2. Principle of the Spark coil (i.e., Ignition Coil).

a. A solenoid consists of 300 turns of wire and it is 6.0 cm in length. A second coil which has 40,000 turns is wound over the same length, and is wound on top of the first coil. (Therefore they have the same axis). The inner coil has an area of 2.50 cm^2. The outer coil has an area of 5.00 cm^2.

The current in the inner solenoid is changed from 0 to 12.0 A in 0.0027 s. What is the final B-field inside the inner coil? What is the voltage induced in the outer solenoid?

b. Same setup as previous, but now the inner solenoid has an iron core. The value of μ for the iron core is 10. If the current in the inner solenoid is again changed from 0 to 12.0 A in 0.0027 s, what is the voltage induced in the outer solenoid? (Hint: In a ferromagnetic material, the applied solenoid B-field is increased by a factor μ, which is called the *relative permeability* of the material)

Answers: 1. 34.0 V   2. a. $B = 7.54 \times 10^{-2}$ T ; emf = 279 V   b. 2790 V

11.2 Motor / Generator

I. Motors: Describe the following essential components of a motor – how are they realized and why are they needed?

A. **Magnetic field**

B. **Coil of wire**

C. **Electric circuit**

D. **Mechanical System**

II. Generators: Describe the following essential components of a generator – how are they realized and what do they contribute to operation?

A. **Magnetic field**

B. **Coil of wire**

C. **Electric circuit**

D. **Mechanical System**

III. Describe the flow of energy and energy conservation in (1) the motor and (2) the generator:

11.2.1 Rail Gun Motor/Generator

A rail gun is a *linear* motor – linear, in the sense that the moving part moves in a straight line, rather than in rotation. The moving projectile is accelerated by the force of a magnetic field on a large current which passes through the projectile. Figure 11.2 is a schematic drawing of a railgun.

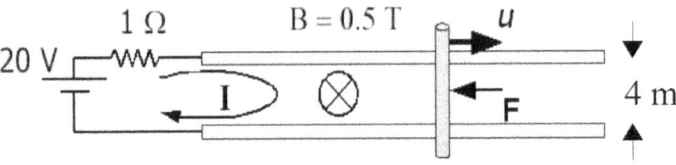

Figure 11.2: Railgun linear motor

Problem: The rail gun is powered by a 20 Volt battery. There is 1 ohm resistance in the circuit. The rail is 4 m wide and moves at various speeds, u, because we are externally applying whatever force is necessary to keep it moving at that speed.

1. Use Faraday's law to determine the back EMF, V_{back}. (Hint: the rate of change of the area $= 4 \times u$)

2. That combines with the battery to give a net voltage in the circuit, V_{net}.

3. Find the external force F needed to balance the magnetic force, so that the rail moves at constant velocity.

4. Figure out the electrical power and the mechanical power. (Hint: the mechanical power is $P_{rail} = F \times u$)

u (m/s) rail vel.	Vback (V) Back EMF	Vnet (V) Net Voltage	I (A) Current	Pbatt (W) Batt. Power	Force on rail, F (N)	Prail (W) Mech Power
0						
2						
5						
8						
10						
12						

Answers: for u = 2 m/s, $V_{net} = 8 \text{ m}^2/\text{s} \times 0.5 \text{ T} = 4.0 \text{ V}$ $P_{batt} = 20 \text{ V} \times 16 \text{ A} = 320 \text{ W}$

Chapter 12

AC Circuits

Alternating Current ('AC') is the common way to generate electric power and distribute it from central power plants to households and factories. AC circuits are also very important in wireless communication. Voltage is applied in one direction and then the opposite direction, and the charges therefore move in a back-and-forth motion.

AC voltage usually has a sinusoidal time dependence, $V = V_{peak} \sin(2\pi f t)$ as illustrated in the following graph:

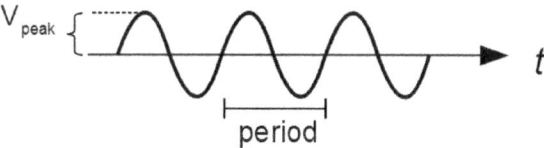

When this voltage is applied to a resistance, a sinusoidal current, $I = V_{peak} \sin(2\pi f t)/R$ results. Therefore the dissipated power is $P = V_{peak}^2/R \, \sin^2(2\pi f t)$. So the power oscillates from a minimum value of 0, when sin is 0, to a peak value of V_{peak}^2/R when sin is ± 1. Average power is $P_{avg} = \frac{1}{2} V_{peak}^2/R$.

It is convenient to define scaled down values of voltage and current in AC circuits. Define:

$$VAC = \frac{V_{peak}}{\sqrt{2}} \qquad \text{defines AC voltage} \qquad (12.1)$$

$$I_{rms} = \frac{I_{peak}}{\sqrt{2}} \qquad \text{defines AC current} \qquad (12.2)$$

$$P = P_{avg} \qquad \text{defines AC power as average power} \qquad (12.3)$$

With these definitions,

$$P = VAC * I_{rms} \qquad \text{AC power} \qquad (12.4)$$

$$VAC = I_{rms} * R \qquad \text{Ohm's Law} \qquad (12.5)$$

1. You have a 230 VAC convection oven. The current is $I_{rms} = 10$ A_{rms}. Find:

$I_{avg} =$ \hspace{2cm} $V_{avg} =$ \hspace{2cm} $P_{avg} =$

$I_{peak} =$ \hspace{2cm} $V_{peak} =$ \hspace{2cm} $P_{peak} =$

$I_{min} =$ \hspace{2cm} $V_{min} =$ \hspace{2cm} $P_{min} =$

R = V_{peak} / I_{peak} =

R = V_{min} / I_{min} =

R = VAC / I_{rms} =

Answers: 1. $I_{avg} = 0$ A $V_{avg} = 0$ V $P_{avg} = 2300$ W $I_{peak} = 14.1$ A_{rms} $V_{peak} = 325$ V $P_{peak} = 4600$ W $I_{min} = -14.1$ A $V_{min} = -325$ V $P_{min} = 0$ R = 23 Ω

2. Consider an old-style, tungsten 100 W light bulb operating on house current, 120 VAC 60 Hz.

a. What is the average current of the lightbulb?

b. What is the average power of the lightbulb?

c. What is the peak power of the lightbulb?

d. What is the peak current of the lightbulb?

e. Is the temperature of the lightbulb filament constant when run on AC power?

3. The power station has oil-filled equipment operating at 230,000 VAC. The oil has a dielectric breakdown strength of 27 Megavolt per meter. What is the minimum spacing of oil between a wire carrying this voltage and a grounded container at 0 volts? (Hint: the spacing must withstand the peak voltage, not just the AC voltage)

Answers: 2. a. 0.833 A   b. 100 W   c. 200 W   d. 1.18 A   3. 1.20 cm

Chapter 13

The Transformer

A transformer is basically two coils that share the same magnetic flux. Transformers are useful for isolating two AC circuits, and especially, for *transforming* the output of a power source so that it properly matches a load. They do this by raising or lowering the voltage output and current capability of the power source to suit the requirements of the load.

Most low frequency transformers make use of an iron core which is shared between the two solenoids. This core enhances the magnetic field and guides it so all the flux in one coil reaches the other coil. Such a transformer is symbolized as shown in the figure:

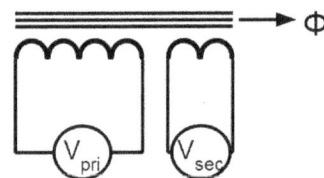

Figure 13.1: Symbol for a transformer with an iron core (indicated by the parallel lines)

Figure 13.1 shows more turns in the primary than the secondary. This is called a 'step-down' transformer, because the secondary voltage is lower than the primary voltage. However, the roles can be reversed, and a step-up transformer produces a secondary voltage that is higher than the primary voltage.

The equation governing voltages is

$$\frac{V_s}{V_p} = \frac{N_s}{N_p} \tag{13.1}$$

Here V_s and V_p are secondary and primary voltages, resp., and N_s and N_p are the secondary and primary coil turns, resp. Because ideally the transformer doesn't consume any power, we must have power in = power out, or $V_s * I_s = V_p * I_p$. That combines with Eq (13.1) to give

$$\frac{I_s}{I_p} = \frac{N_p}{N_s} \tag{13.2}$$

Equation (13.2) shows that while the voltage ratio is proportional to the turns ratio in the transformer, the current ratio is *inversely* proportional to the turns ratio.

1. In your neighborhood, a transformer on top of a utility pole, or one connected to underground wires, transforms 7,200 VAC from the local substation into 225 VAC for use in your home.

a. What is the turns ratio of this step down transformer?

b. If you are using 50 A in your home to operate appliances including an air conditioner, how much current is coming from the substation to supply the transformer for your house?

2. You travel to the Far East where the AC voltage is 220 VAC 50 Hz. You operate your U.S. hair dryer using a transformer to step down the voltage to 110 VAC.

a. What is the turns ratio (primary-to-secondary)?

b. The hair dryer is designed to deliver 1675 Watts. What is the current in the secondary?

c. What is the current in the primary?

d. What is the power delivered to the primary?

e. With all that power going into the primary, why doesn't the transformer get very hot? What is the minimum required rating of the transformer in Volt-Amperes (VA)?

Answers: 1. a. 32:1   b. 1.56 A   2. a. 2:1   b. 15.2 A   c. 7.61 A   d. 1675 W

3. A small transformer for neon signs is designed to plug into the wall (house current = 120 VAC) and supply an output voltage of 5000 VAC to a window sign.

a. Is this a step-up or step-down transformer (circle one) ?

b. If there are 300 turns on the primary, how many turns are there on the secondary coil?

c. The window sign uses a power of only about 20 W. What is the current in the secondary?

d. What is the peak voltage being applied to the neon sign?

e. What is the minimum instantaneous voltage applied to the neon sign?

f. What is the peak power used by the neon sign?

g. What is the rms current in the primary?

Answers: 3. a. step-up b. 12,500 c. 0.004 A_{rms} d. 7070 V e. -7070 V f. 40 W g. 0.167 A_{rms}

Efficiency: 4. A step-down transformer is employed in the power supply for a piece of electronic audio equipment.

a. When the transformer is working near full rated capacity, the input is 120 VAC and 1.35 A_{rms}. Some power is lost in the iron core of the transformer, so the efficiency is 92%. If the output voltage is 12 VAC, What is the output current?

b. When the transformer is working at low power, there is still a significant loss of energy, and the efficiency drops to 65%. A typical situation is that the input voltage is still 120 VAC, but the input current is only 0.135 A_{rms}. The output voltage is still 12 VAC. What is the output current?

Resistance Matching: 5. A loudspeaker has a 3.2 Ω input *impedance* (similar to resistance, but 'impedance' is used in many signal applications). The audio driver circuit would function best if it could 'see' an impedance of 320 Ω. The audio driver delivers a voltage of 100 VAC to the primary of a matching transformer.

a. Use Ohm's Law to find the primary current, based upon 320 Ω primary impedance.

b. If the secondary voltage of the matching transformer is 10 VAC, what is the turns ratio?

c. What is the secondary current, if the power is transmitted from the primary with 100% efficiency?

d. Show using Ohm's law again that the ratio between secondary voltage and secondary current agrees with the impedance of the loudspeaker, i.e., 3.2 Ω.

Answers: 4. a. 12.4 A_{rms} b. 0.878 A_{rms} 5. a. 0.313 A_{rms} b. 10:1 c. 3.13 A_{rms}

13.1 power transmission

1. A power station several miles from your home generates electricity. There is a step-up transformer at the power station, and a step-down transformer near your house. See figure 13.2

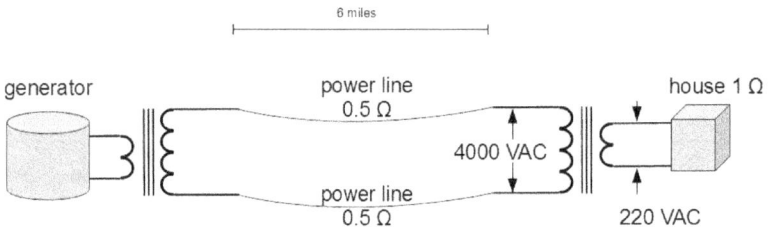

Figure 13.2: Power to your house uses transformers

Figure out how much power is lost in the two 0.5 Ω power lines as follows:

a. What is the turns ratio in the transformer in your neighborhood, by your house?

b. What is the current used by your house in this example?

c. What is the primary current in the transformer that feeds your house?

d. How much power is lost in the power lines?

e. If the transformers weren't there, and you used the same amount of current in your house, how much power would be lost in the power lines?

Answers: 1. a. 18.2:1   b. 220 A_{rms} c. 12.1 A_{rms} d. 146 W   e. 4.84×10⁴ W

13.2 Quiz on AC Transformers

Question 1. A small transformer is used to power a clock-radio in your bedroom. The clock-radio operates at a voltage of 7.5 VAC, and uses 6.0 W of power. Assume the primary plugs into the wall outlet, which provides 120 VAC to the primary of the transformer.

a. If there are 450 turns of wire in the primary, what is the number of turns in the secondary?

$N_s =$ _____ turns

b. What are the primary and secondary currents?

$I_p =$ _____ A_{rms}

$I_s =$ _____ A_{rms}

c. What are the primary and secondary power, assuming 100% transformer efficiency?

$P_{sec} =$ _____ W

$P_{pri} =$ _____ W

13.2.1 Answers to Quiz on AC Transformers

a. The turns ratio is the same as the voltage ratio, so $N_s/N_p = V_s/V_p$ = 7.5 VAC / 120 VAC = 0.0625 . Therefore N_s = 450 * 0.0625 = 28.1 .

b. Use the information about power, 6.0 W, and the primary and secondary voltages to determine the primary and secondary currents: I_p = 6 W / 120 VAC = 0.050 A_{rms} , and I_s = 6 W / 7.5 VAC = 0.80 A_{rms} .

c. Assuming 100 % efficiency, secondary power is equal to primary power, 6.0 W.

Note: If the problem specified a lower transformer efficiency, say 90 %, then the primary power would have to be higher, 6.0 W / 0.9 = 6.67 W, in order to meet the power requirement of 6.0 W in the secondary. This would require a proportionately greater primary current in part b). Turns ratio and secondary properties would remain the same.

Chapter 14

Traveling Waves

Wave motion is highly important for the transmission of information and energy over a distance. Figure 14.1 shows a portion of a wave traveling to the right, and the same portion one *period* later.

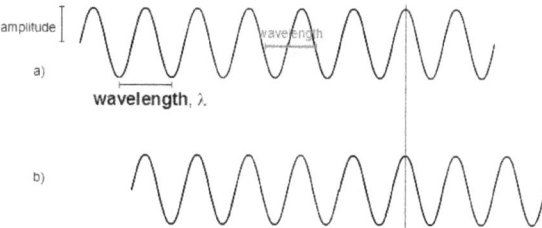

Figure 14.1: a. Snapshot of a wave at one instant of time. b. Same wave, one period later.

The figure indicates that the wave group moves exactly 1 wavelength in 1 period. Therefore, the speed of the wave, often denoted by 'c', is

$$c = \frac{\lambda}{T} \qquad (14.1)$$

where λ is the wavelength, in meteres, and T is the period, in seconds. The frequency is just f = 1/T, so we also have

$$c = f\lambda \qquad (14.2)$$

1. A circus performer on a trapeze goes back and forth every 3 seconds. What are the period and frequency of the motion?

Answers: 1. T = 3 s, f = 0.333 Hz

2. What is the period of a 440 Hz sound ('A')? If the velocity of sound in air is 330 m/s, what is the wavelength?

3. A honey bee beats its wings 660 times each second. What is the frequency of this sound, and what is the wavelength?

4. When an automobile is traveling toward you the horn seems to have a higher pitch, whereas when traveling away, the frequency seems lower. (Doppler effect) Why?

5. A snapshot shows the crests of a water wave are 5 meter apart. If you are standing on a rock and observe the portion of a wave between two crests passes by in 3 seconds, what is the wave speed?

Answers: 2. 0.75 m   3. 660 Hz   0.5 m   5. 1.67 m/s

14.1 Electromagnetic waves and photons

In a vacuum the speed of all electromagnetic waves is denoted by c = 3.00×10^8 m/s. The speed of light in air is virtually the same as in vacuum.

1. Red light has a wavelength of 700 nm. What is the frequency?

2. Blue light has a wavelength of 400 nm. What is the frequency?

3. A photon of light is the smallest possible energy unit. The energy of a photon is given by

$$E_{photon} = h\nu \quad \text{where Plank's constant h} = 6.63 \times 10^{-34} \text{J-s} \quad (14.3)$$

The above equation customarily uses the Greek letter ν for frequency. What are the energies of a red and a blue photon?

4. What are the wavelength and energy of microwave photons (frequency = 2.45 GHz) ?

5. What are the wavelength and frequency of medical x-rays (typical x-ray photon energy $\sim 3 \times 10^{-15}$ J)?

Answers: 1. 4.29×10^{14} Hz   3. red: 2.84×10^{-19} J   5. $\nu = 4.52 \times 10^{18}$ Hz $\lambda = 6.63 \times 10^{-11}$ m

14.2 Light sources - Inverse square law

1. a. The bulb on the right is 25× brighter than the bulb on the left, when viewed from the same distance.

(The picture below is a birds-eye view.)

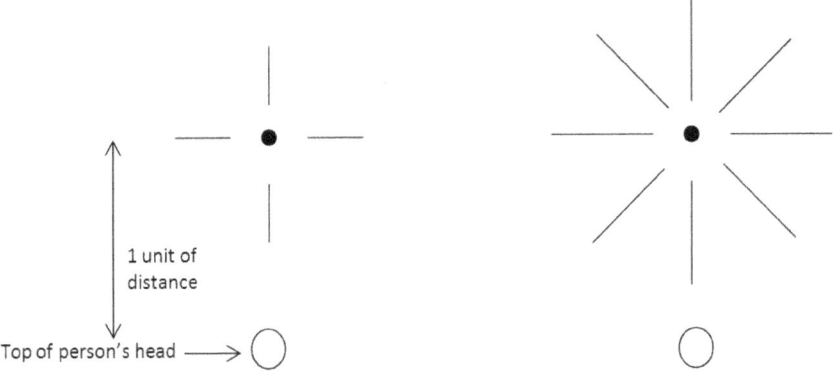

b. How far back would the light on the right have to be moved back so that the lights would be seen equally bright to the person?

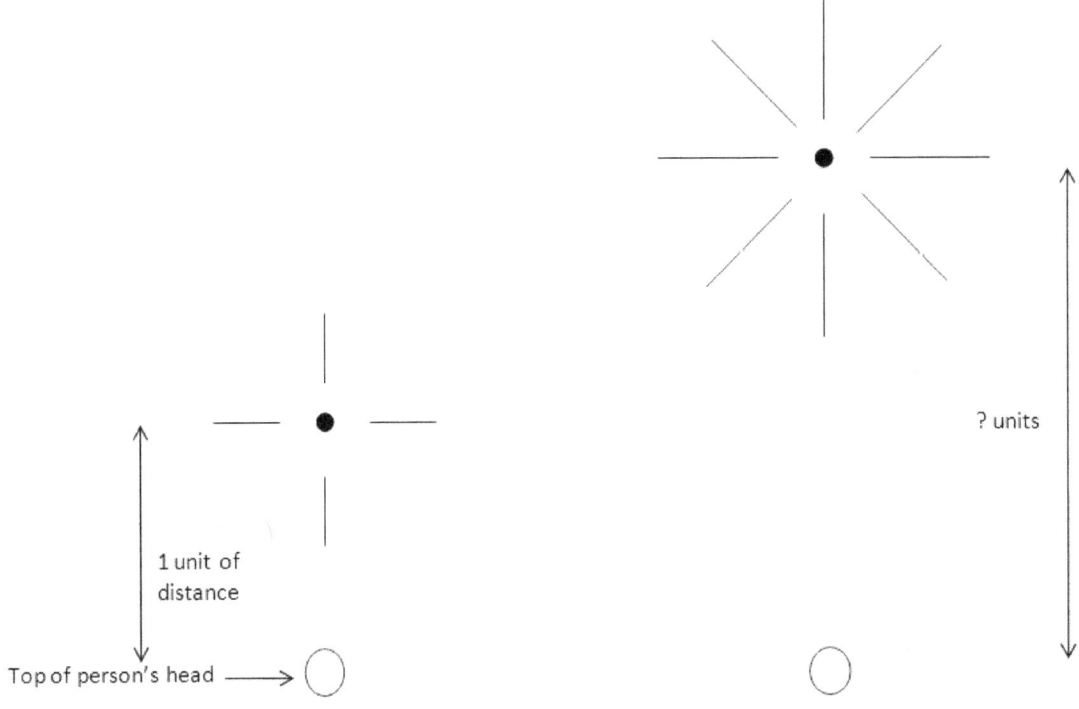

Answers: 1. b. 5×

14.2. LIGHT SOURCES - INVERSE SQUARE LAW

2. The light on the right is 20 times further away from the person than the light on the left. How many times brighter must the light on the right be so that the lights appear equally bright to the person?

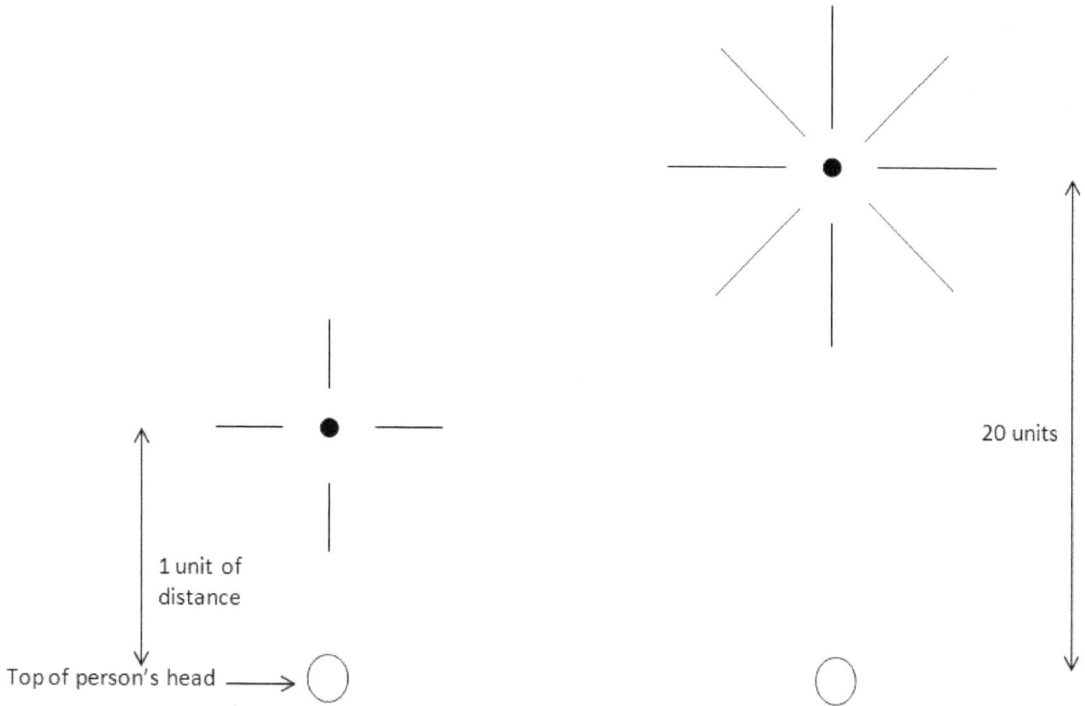

Answers: 2. 400

14.2.1 Light intensity

One measure of light intensity is Watts per square meter. 3. Measure the brightness of a light bulb of known power and at a known distance. Then measure and compare the brightness of the sun, with unknown power P_{sun} at a known distance of 150 million km.

a. Use the lightmeter to read the energy intensity of the sun. Point the detector cylinder directly at the sun.

Intensity reading of sun _____

b. Now point the sensor at a clear glass lightbulb. Adjust the distance between sensor and lightbulb so that it reads the same as it did in part a.

Distance to center of lightbulb _____

Light energy intensity is the same in a. and b. But the distance from the sun is much, much bigger. Determine the square of the distance ratio. Use $R_{sun} = 1.50 \times 10^{11}$ m.

Distance ratio squared _____

Intensity ratio = _____

If the light bulb emits P Watt, how many watts does the sun emit?

Suns Watt output P_{sun} = _____

14.2. LIGHT SOURCES - INVERSE SQUARE LAW

4. The total power emission of the sun is 3.85×10^{26} watts, in various wavelengths.

a. How much energy/second reaches a square meter of Arizona sidewalk? (Hint: find the number of square meters on the surface of a sphere whose radius is the radius of the Earth's orbit: $R_{orbit} = 1.50 \times 10^{11}$ m, Area $= 4\pi R_{orbit}^2$)

Intensity = _____ W/m^2

b. Given the answer in part a), which is the number of joules reaching a square meter of Earth in 1 second, what is the number of photons reaching a square meter in 1 second? (You can use the energy of a green photon, starting from a wavelength for green light = 550 nm)

Photons /sec/m^2 = _____

c. A green laser pointer has a power of 0.002 W. The power is directed in a beam that is only 4 mm^2. What is the number of photons emitted per second by the laser?

Photons /sec _____

d. How does the answer in c) compare with the number of photons/sec which is barely detected by the dark-adapted human eye ($\simeq$ 100 photons/sec)?

e. How does the answer in c) compare with the intensity of sunlight at the Earth's surface? (Hint: Use the laser power and scale it up from 4 mm^2 to 1 m^2)

Answers: 3. a. 1.36×10^3 W/m^2 b. 3.77×10^{21} c. 5.53×10^{15} d. 500 W/m^2

14.3 Refraction and Snell's Law

When a light beam in air strikes a smooth dielectric surface at an oblique angle, there is reflection and refraction. The angle of reflection equals the angle of incidence. What is *refraction*? Light entering a medium with slower speed of light *always* bends *toward* the normal. When light crosses into a faster medium, it bends away from the normal. The angle of refraction is given by Snell's Law:

$$n_i \sin \theta_i = n_r \sin \theta_r \qquad \text{Snell's Law} \tag{14.4}$$

where n_i and n_r are the refractive indices of the incident and refracting media, resp., and θ_i and θ_r are the corresponding angles of incidence and refraction.

1. A light ray traveling in air strikes the surface of a slab of glass at an angle of incidence of 50° Part of the light is reflected and part is refracted. Find the angles the reflected and refracted rays make with respect to the normal to the air-glass interface. The index of refraction for glass is 1.50.

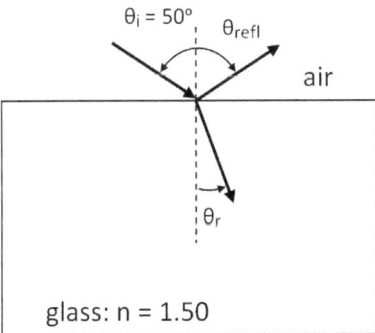

2. A ray of light crosses the boundary from glass to air. If the angle of incidence is 20°, what is the angle of refraction?

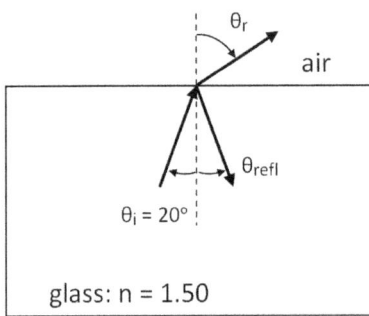

Answers: 1. $\theta_{refl} = 50°$ $\theta_r = 30.7°$ 2. $\theta_{refl} = 20°$ $\theta_r = 30.9°$

14.3. REFRACTION AND SNELL'S LAW

3. A ray of light crosses the boundary from glass to air. If the angle of incidence is 41°, what is the angle of refraction? Draw the reflected and refracted rays, indicating the angles and showing angles in correct proportion.

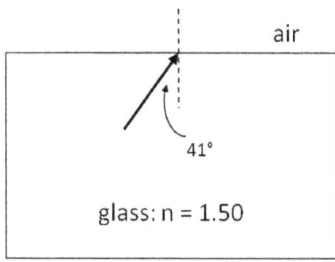

4. A ray of light crosses the boundary from glass to air. If the angle of incidence is 42°, what is the angle of refraction? Draw the outgoing ray(s), indicating the angle(s) and showing the angle(s) in correct proportion.

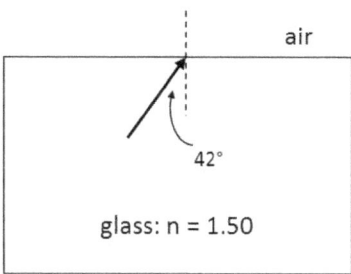

Answers: 3. $\theta_{refl} = 41°$ $\theta_r = 79.8°$ 4. $\theta_{refl} = 42°$ no refracted ray - total internal reflection

5. A ray of light crosses the boundary from diamond to air. If the angle of incidence is 24°, what is the angle of refraction? Draw refracted and reflected rays.

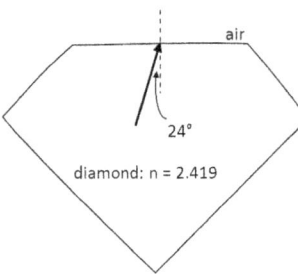

6. A ray of light crosses the boundary from diamond to air. If the angle of incidence is *now* 25°, what is the angle of refraction? Draw refracted and reflected rays.

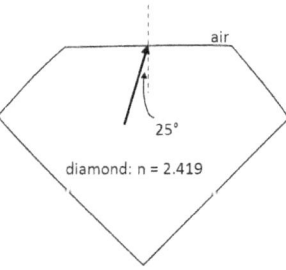

7. Using trial and error with your calculator, determine what is the exact minimum angle of incidence giving TIR (total internal reflection) in diamond? (You can also get this result from Equation (14.4) assuming $\theta_r = 90°$)

Answers: 5. $\theta_{refl} = 24°$ $\theta_r = 79.8°$ 6. $\theta_{refl} = 25°$ no refraction - total internal reflection 7. $\theta_i = 24.418°$

8. **Mirage:** The refractive index of air is not exactly 1, its really 1.000292 at room temperature. However, at the surface of a hot road, the air is less dense and the refractive index is 1.000245. Consider the air above the road as a sandwich of hot and cool air. If a light ray approaches the boundary between the hot and cool layers at a large angle of incidence, what is the critical angle at which there will be total internal reflection? What is your experience with this type of reflection?

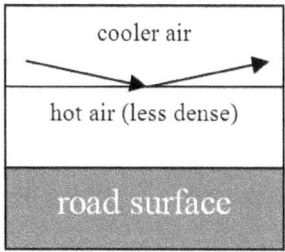

Figure 14.2: A mirage due to the low-index air layer at the hot road surface

Answers: 7. $\theta_{TIR} = 89.44°$

8. **Lightpipe:** A lightpipe, depicted below, accepts light from a wide range of incident angles, and 'pipes' it very far to the opposite end without any light leaking out. The refracted rays entering the pipe hit the wall of the light pipe at a glancing angle and total internal reflection results.

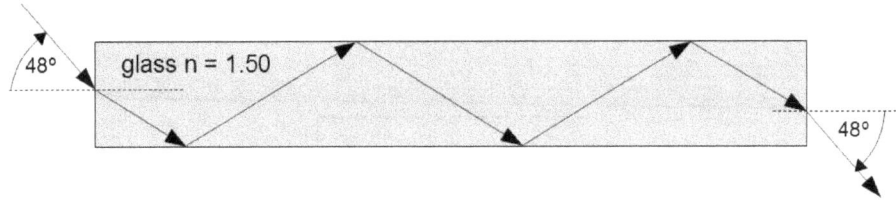

a. Find the angle of refraction for a ray striking the lefthand glass surface at an incident angle $\theta_i = 48°$.

b. Show that the refracted ray found in part a) above suffers total internal reflection at the bottom glass-air interface. Hint: the angle of incidence at the bottom glass-air interface is the complement of the angle found in part a) above. What is this angle of incidence?

c. What is the critical angle for TIR for a ray traveling in glass incident on the glass-air interface?

d. Find the maximum angle of incidence on the lefthand air-glass interface that will result in total internal reflection at the bottom glass-air interface.

Answers: 8 a. $\theta_r = 29.7°$ b. $\theta_i = 60.3°$ c. $\theta_{TIR} = 41.8°$ d. $90°$ – i.e., all angles are piped

14.3. REFRACTION AND SNELL'S LAW

9. **Underwater:** What do you (or fishes) see when you are underwater and look upwards? If you look straight up, you see the sky, the sun, the birds in the sky. If you look at an angle upwards, the surface of the water reflects whatever is on the bottom of the pool. At a certain angle, you see whatever is sitting at the edge of the pool.

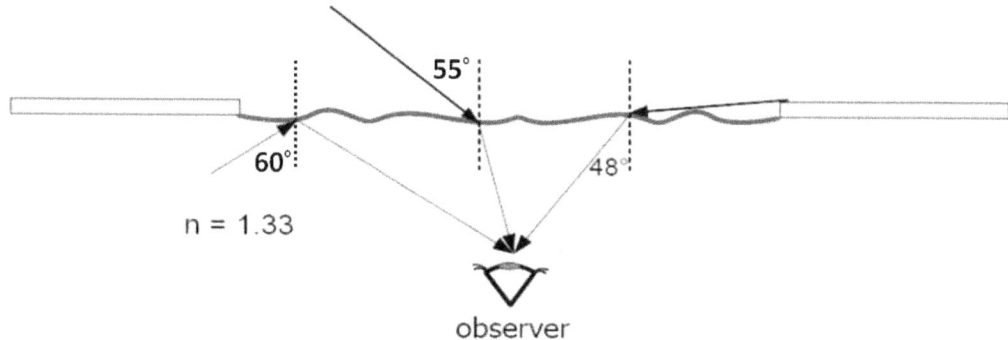

For each of the rays in the above figure of a swimming pool in cross section, calculate and fill in the value of the corresponding reflected, refracted, or incident angle. Start with

a. What happens to the ray approaching the surface from under the water at a 60° angle, and what does the observer see?

b. What happens to the ray approaching the surface from above at a 55° angle, and what does the observer see?

c. When a ray approaches the observer from a 48° angle, from what glancing angle did it reach the water, and what does the observer see?

Answers: 9 a. reflects the bottom: $\theta_{refl} = 60°$ b. bends to $\theta_r = 38.0°$ c. sees the edge of the pool: $\theta_i = 81.3°$

Chapter 15

Wave interference

When two traveling waves cross, the energy and information they carry pass right through each other. But the medium undergoes motion which is a combined effect of the two waves. This is called *interference*.

1. **Constructive interference:** Two transverse square waves approach from opposite directions. One is depicted as a solid line, and one is dotted. When they cross, starting at t = 4, they interfere. At each time, use simple addition in the y-direction and draw a single line shape to show how the combined wave appears.

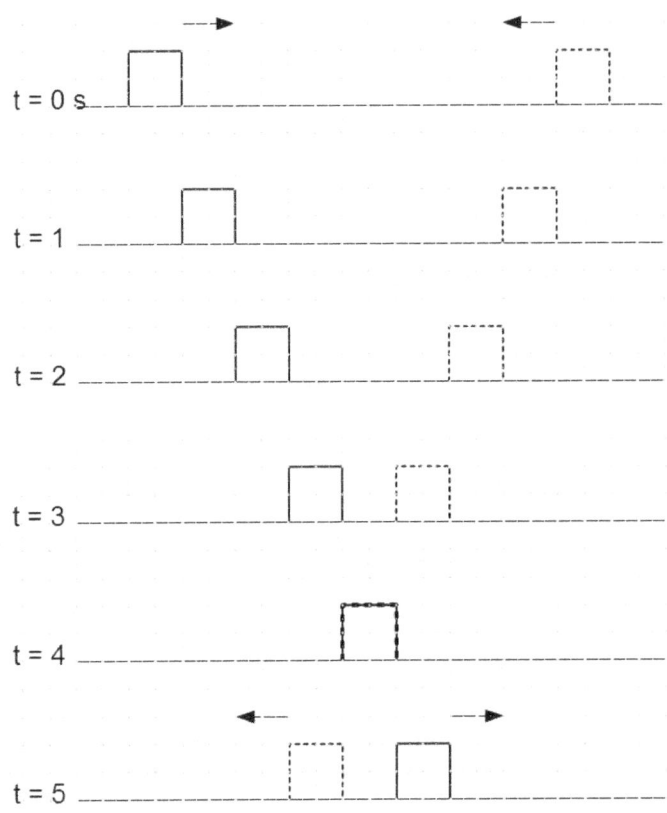

2. **Destructive interference:** a. Transverse square waves of opposite sign approach from opposite directions. Use simple addition in the y-direction and draw a single line shape to show how the combined wave appears.

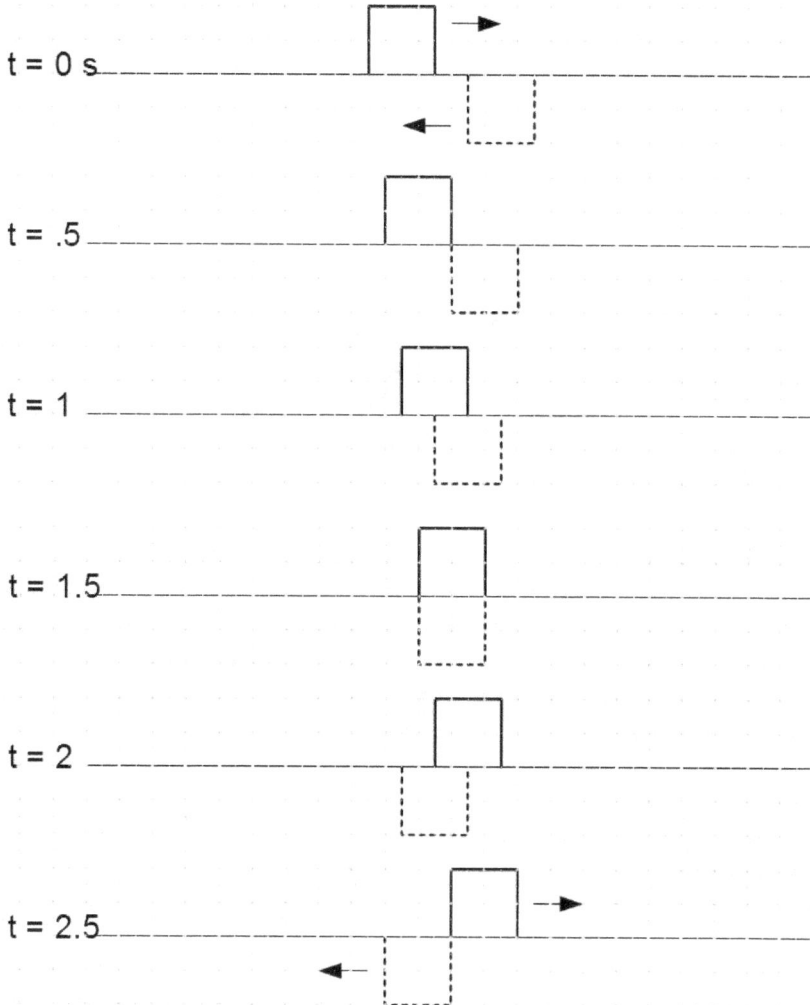

b. At t = 1.5s, the waves momentarily cancel completely. Where has the wave energy gone that was carried by the two waves?

3. **Standing Wave:** Two waves having the same frequency approach from opposite directions. Add the waves in the y-direction and draw a single combined wave. Why is this interference pattern called a 'standing' wave?

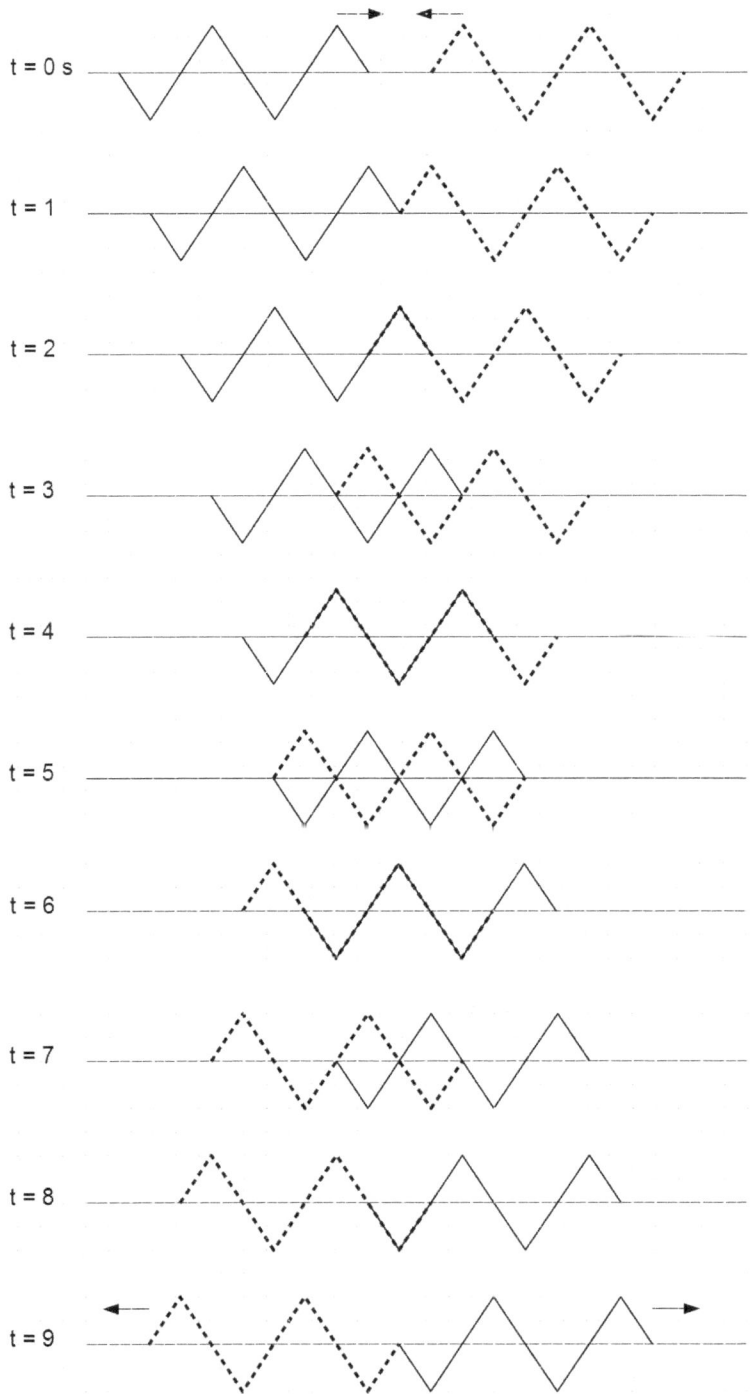

15.1 Wave Interference from 2 coherent sources

An electric guitar is connected to two separate speakers, so that the vibrations of the two speakers are identical. Identical waves emanate from the two speakers.

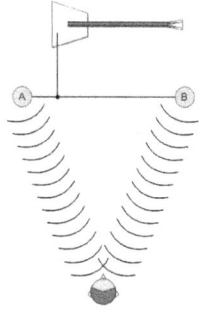

1. The listener in the above figure is positioned equidistant from the two speakers. The sound wavefronts reaching the listener arrive together 'in phase' and add constructively. The sound is then louder than either speaker alone.

However, if the listener is positioned to the side, as shown in figures 15.2 and 15.1, the sound waves from the two speakers will interfere, as indicated by the figures. Constructive or destructive intereference will depend which note the guitar is playing, because different notes have different frequencies and different wavelengths.

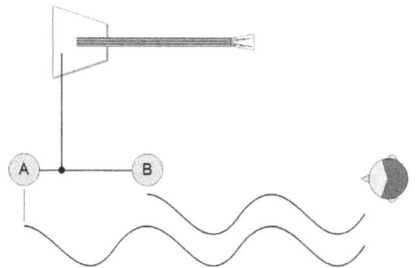

Figure 15.1: Constructive interference

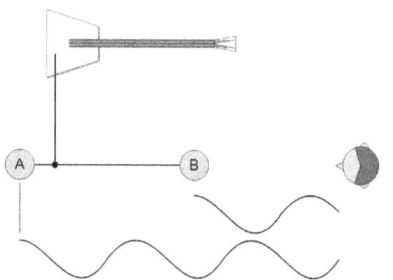

Figure 15.2: Destructive interference

1. If the guitar plays an 'A' (440 Hz), and the speed of sound is 330 m/s, what is the horizontal separation between the speakers in each of the two figures, 15.1 and 15.2?

2. In figure 15.1, assuming the note played is 'A' again, what are the next two larger speaker separations that give constructive interference (maximum loudness)?

Answers: 1. 0.75 m   1.25 m   2. 1.5 m, 2.25 m

Two speakers are spaced d = 1.0 meter apart. The both are playing the same 1100 Hz tone.

1. What is the wavelength of the tone, assuming the speed of sound is 330 m/s?

2. The observer is standing far away and at an angle θ away from the forward direction of the speakers. See figure below. The first $\theta > 0$ at which there is an interference maximum of the sound is depicted in the figure.

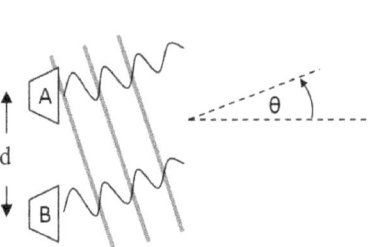

The expanded figure below shows a triangle formed by a line between the two speakers, the angle θ, and the wavelength λ. Find the angle θ using the equation, $\lambda = d \sin \theta$.

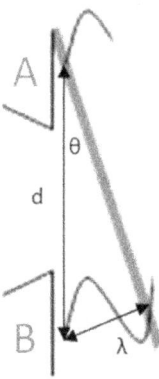

3. When θ is negative (observer moves downward in the figure) does $\lambda = -d \sin \theta_{-1}$ also give an interference maximum? If so, at what angle θ?

4. What if the angle θ is made larger (listener moves more sideways to the speakers), and 2 wavelengths can fit in the short leg of the triangle in the above figure? Find θ when $2\lambda = \pm d \sin \theta_{\pm 2}$. Find θ_2, θ_3, and θ_4; and θ_{-2}, θ_{-3}, and θ_{-4}.

Answers: 1. 0.30 m   2. 17.5°   3. −17.5°   4. $\theta_{\pm 2} = \pm 36.9°$ $\theta_{\pm 3} = \pm 64.2°$ $\theta_{\pm 4}$ not possible

15.2 Grating diffraction of light

Multiple light beams can be produced from any light source by interposing a grating. Interference between light waves is called *diffraction*.

Three of the openings in a larger grating are illustrated below:

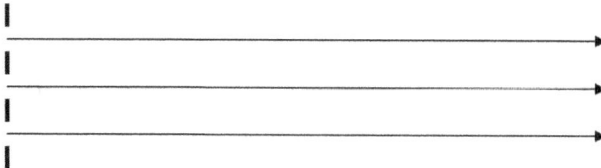

In the forward direction (along $\theta = 0$, the interference is constructive. Different paths to the observer are all the same length. This is where n = 0, in the formula, $n\lambda = d\sin\theta$. The forward beam is called the principle maximum.

At an angle along θ such that the wavelength, $\lambda = d\sin\theta$, the interference is also constructive, because each path is one wavelength longer than the next. This is called the n = 1 fringe, and is represented in the figure below.

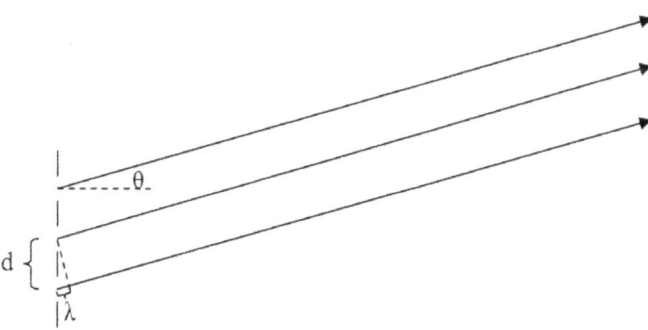

1. Red light has a wavelength of 630 nm. If the slits in the grating are spaced 1500 nm apart, what will be the diffraction angle for the first diffraction maximum?

2. Blue light has a wavelength of 440 nm. If the slits in the gating are still 1500 nm apart, what will be the diffraction angle for the first diffraction maximum for blue light?

Answers: 1. 24.8° 2. 17.1°

Question. A diffraction grating has multiple openings separated by 2.50×10^{-6} m. It is illuminated by red light having a wavelength $\lambda = 632$ nm. What are the angles θ_1 and θ_2 of the 1st and 2nd order interference maxima from this grating?

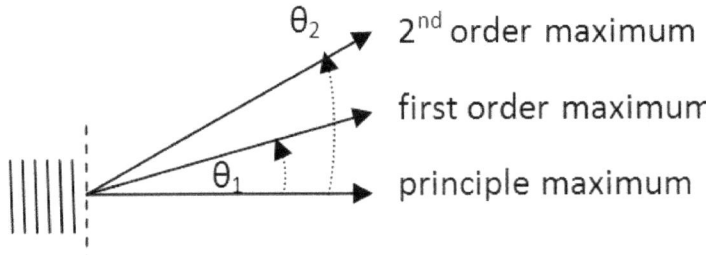

$\theta_1 =$

$\theta_2 =$

15.2. GRATING DIFFRACTION OF LIGHT

3. Grating spacing is often expressed indirectly in terms of the number of lines per millimeter. A grating has 550 lines per mm.

a. What is the spacing of the slits in this grating (convert your answer to nm)?

b. In this grating, what will be the angles of the first diffraction maxima for red light (630 nm) and for blue light (440 nm)?

c. As the spacing between slits in the grating increases, the separation between diffraction maxima increases or decreases?

d. As the wavelength decreases, the diffraction angle increases or decreases?

Answers: 3. a. 1820 nm   b. red: 20.3°   blue: 14.0°

15.2.1 Diffraction in nature

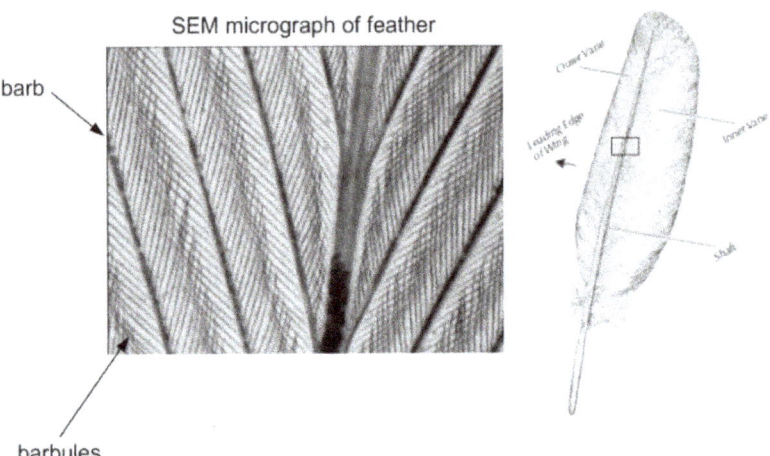

Feather diffraction A feather is made up of little ribs, called 'barbs', that are evenly spaced and 0.3 mm apart. This repeating pattern can diffract light.

1. a. If a beam of light having wavelength 540 nm impinges on a feather, what is the angle of the first, second and third diffraction maxima?

b. Each of the barbs described in part a) contain *barbules*, which are smaller structures arranged like the teeth of a comb along the barb. These also will diffract the light. Using 540 nm wavelengh light, the barbules produce a first order diffraction maximum at 1.96°. What is the separation, **d**, in millimeters, between the barbules?

Atomic spectra - Grating spectroscope 2. The sodium element spectrum has a bright yellow double line at 590 nm. If a grating is used to separate the spectral colors, and the grating has 550 lines per mm, what is the angle of the 1st maximum of the sodium spectrum?

Answers: 1. a. 0.103° 0.206° 0.309° b. 0.0158 mm 2. 18.9°

15.2. GRATING DIFFRACTION OF LIGHT

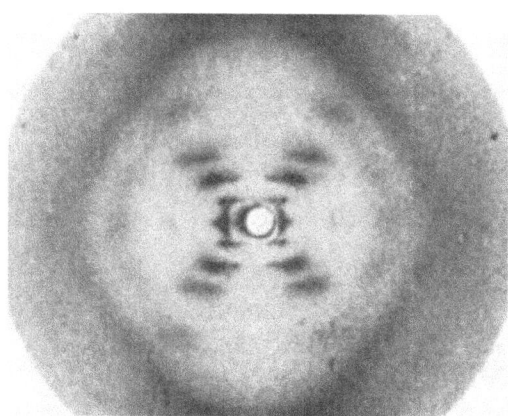

Figure 15.3: Famous Image 51 – Rosalind Franklin's x-ray diffraction picture of DNA, which makes the DNA helix structure evident

X-ray diffraction X-ray diffraction was the tool which Rosalind Franklin used to discover the double helix in DNA. DNA has two helices intertwined. It is a 'double helix' (See Figure 15.4).

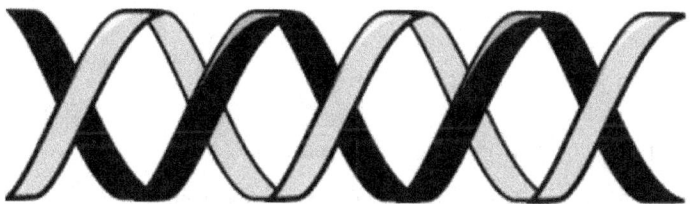

Figure 15.4: The x-ray waves 'see' two gratings – the front one slanting from lower left to upper right, the back one (behind) slanting from upper left to lower right

3. a. Figure 15.4 shows how the double helix, as viewed from the side, looks like two crossed gratings. Consider just one grating out of the two. The helix has a half-pitch (spacing between turns) equal to 1.41 nm. If the x-rays in a diffractometer have a wavelength $\lambda = 0.16$ Å, what is the angle of diffraction of the first maximum in a DNA diffraction pattern? [1 Å = 0.1 nm].

b. The repeating spacing between the phosphate groups along the DNA helix is 0.34 nm. What is the first order diffraction angle for this 'grating'? (Assume the same x-ray wavelength $\lambda = 0.16$)

Answers: 3. a. $0.650°$ b. $2.70°$

Chapter 16

Resonance

In Chapter 15, two waves traveling in opposite directions interfere to produce a standing wave. The standing wave is a **resonance** similar to the resonance of a mass on a spring. That is, energy is alternately stored as potential energy, then released as kinetic energy, then returned back into P.E., ... etc.

In a bounded container of liquid or gaseous material, or in a piece of solid, waves rebound from the boundaries and interfere to produce standing waves. The sinusoidal motion has a definite resonance frequency, and a corresponding wavelength.

1. Figure 16.1 shows three standing waves on a string stretched between two fixed boundaries. The wavespeed on a string of this mass/length and this force of tension is 500 m/s. If L = 0.6 m, what are the frequencies of the three standing wave resonances?

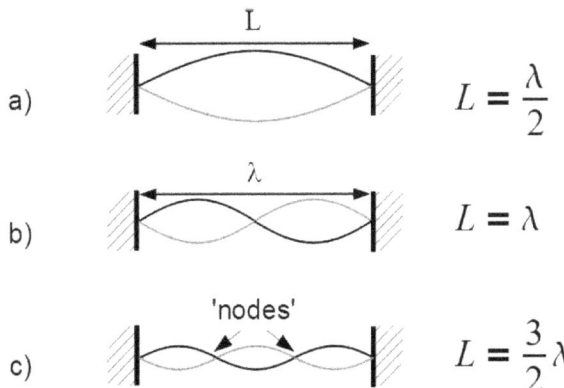

Figure 16.1: String resonances: a) Fundamental, b) 2nd harmonic, and c) 3rd harmonic

a. One half-wave of the fundamental (lowest frequency resonance) fits between the boundaries. Therefore $\lambda = 1.2$ m, and $f_1 = c/\lambda = 417$ Hz.   b. 2nd harmonic: $\lambda = 0.6$ m, and $f_2 = c/\lambda = 833$ Hz.   c. 3rd harmonic: $\lambda = 0.4$ m, and $f_3 = c/\lambda = 1250$ Hz.

2. What is a general rule which relates the wavelength, λ, and the distance between consecutive nodes?

3. What is a general rule for resonances on a string that gives the frequency of the nth harmonic, f_n, in terms of n?

4. The fourth and fifth harmonics of a musical instrument string are 480 Hz and 600 Hz, respectively. The length of the string is 0.4 m.

a. What is the fundamental frequency?

b. What is the speed of waves on this string?

c. When it vibrates at 600 Hz, what is the distance between consecutive nodes on the string?

d. What is the wavelength when the string vibrates at 600 Hz?

Answers: 2. consecutive nodes are λ/2 apart 3. $f_n = 2n/L$ 4. a. 120 Hz b. 96 m/s c. 0.08 m d. 0.16 m

16.1 Sound resonances

Typically sound resonances are set up in tubes or chambers filled with air. Sound is a longitudinal pressure wave. Compression waves are also possible in solids and liquids.

Sound resonances are a little different from waves on a string or chain because the container for sound resonances can have either open or closed ends. At a closed end, the air motion must be zero. At the open end of a tube, the air motion is maximum.

Two Closed ends 1. What are the resonant frequencies for a completely closed pipe having length 10 m? (Assume the velocity of sound is 330 m/s)

Answer: As shown in the figure below, the longitudinal motion of the air in a closed pipe is similar to the transverse motion of string or chain under tension between fixed boundaries.

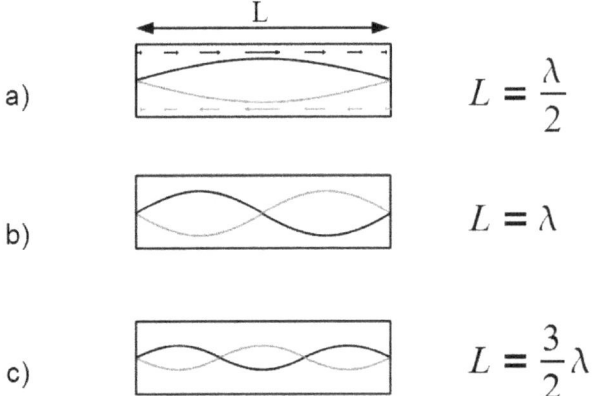

Figure 16.2: Closed pipe resonances: a) Fundamental, b) 2nd harmonic, and c) 3rd harmonic

The arrows indicate that the air moves to the right for one-half cycle, then moves to the left for the next half cycle. There is first compression against the right-hand boundary, then compression against the left-hand boundary.

Figures 16.2 a-c indicate the resonant frequencies are given by the same formula as the string resonances: $f_n = 2nc_{sound}/L$, i.e., $f_1 = 330/20 = 165$ Hz, $f_2 = 330$ Hz, $f_3 = 495$ Hz.

16.1. SOUND RESONANCES

One open end 2. If one end of the pipe is open, air is free to move in and out of the open end.

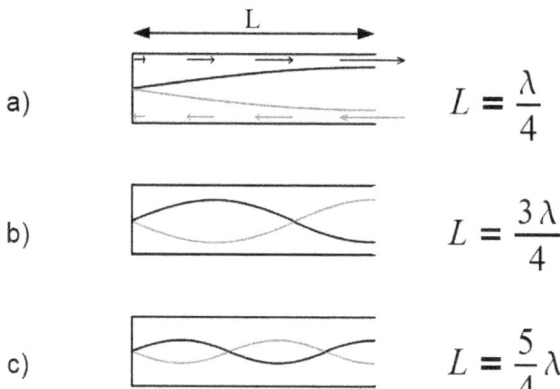

Figure 16.3: One end open resonances: a) Fundamental, b) 2nd harmonic, and c) 3rd harmonic

a. What are the resonant frequencies for a pipe having one end open and length 10 m? (Assume the velocity of sound is 330 m/s)

Hint: Fill in the blanks in the following table. Using information in the first column, solve for λ in the second column. Then use $f = c/\lambda$ for the third column.

L	wavelength, λ	resonant frequency f
$= \frac{\lambda}{4}$		
$= \frac{3\lambda}{4}$		
$= \frac{5\lambda}{4}$		

b. Write a general formula for the frequencies, f_n, of sound resonances in a pipe that is open at one end. Recall that f_1 is the frequency of the fundamental, f_2 is the frequency of the 2nd harmonic, ...

Answers: a. fundamental: $\lambda = 4L$ b. $f_n = (2n-1)c/4L$

Both ends open 3. If both ends of the pipe are open, air is free to move in and out of each end.

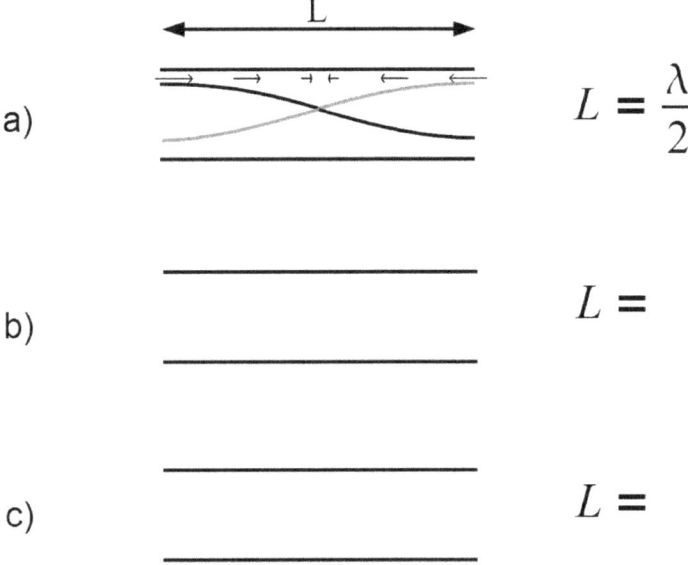

Figure 16.4: Both ends open : a) Fundamental, b) 2nd harmonic, and c) 3rd harmonic

In the figure above, drawing a) shows a graph of how air moves in a pipe open at both ends. The small arrows indicate that the fundamental is a 'breathing' mode, where air moves in and out of the two ends in opposition, so that the pipe 'breathes'.

a. In figures 16.4 b) and c), draw the graph of air motion for the 2nd and 3rd harmonics, resp.

b. In figures 16.4 b) and c), write in the relationship between L and wavelength λ for the 2nd and 3rd harmonics, resp.

c. What are the resonant frequencies for a pipe having both ends open and length 10 m? (Assume the velocity of sound is 330 m/s)

d. Write a general formula for the frequencies, f_n, of sound resonances in a pipe that is open at both ends.

Answers: b. $L = \lambda$, $L = 3\lambda/2$ c. 16.5 Hz, 33 Hz, 49.5 Hz, ...   d. $f_n = nc/2L$

16.2 Other resonance examples

1. A strong wind blows the powerline between two poles, and sets up a standing wave on the wire, as indicated in the drawing. The distance between the poles is 200 m.

a. What is the wavelength of this standing wave?

b. If the frequency of this oscillation is 0.90 Hz, what is the wave speed of transverse waves on the power line?

2. A tuning fork with resonant frequency 2300 Hz is held up to a tube open at both ends. The tube, which has a length 37 cm, strongly enhances the sound of the tuning fork – i.e., a sound wave resonance is excited in the tube. If the velocity of sound is 340 m/s, which harmonic has been excited in the tube?

3. A shower stall is essentially a tube open at both ends, and long enough to cover most of your body. If you sing a C-note at a frequency of 128 Hz, the stall resonates loudly. What is the length of the shower stall?

The Bohr atom: Niels Bohr modeled the hydrogen atom in order to explain the spectral lines of hydrogen. In his model, the electron orbits around the proton, subject to the Coulomb force. The only way to get narrow light emission lines was to posit that the electrons are restricted to a set of discrete orbits at distinct radii, r_n.

According to quantum theory, electrons are particles but they also have wave properties. Considered as a wave, the electron in hydrogen must fit an integer number of wavelengths around a circular orbit. This is similar to a sound resonance in a tube. See figure 16.5. Because the wavelength $\lambda = h/(mv)$, where m*v is the momentum of the electron, each resonant wavelength determines a specific energy level.

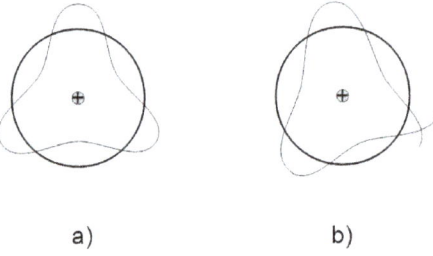

Figure 16.5: a) Bohr hydrogen atom with electron orbit n = 3 b) slightly larger wavelength is not a 'resonant' electron wavelength

The restriction on possible wavelengths, together with the relationship between wavelength and particle momentum, give a relationship between the orbit radius, r_n, and the velocity, v_n in the nth-orbit:

$$n\lambda = 2\pi r_n \quad \text{combine 'resonance' condition on } \lambda \tag{16.1}$$

$$\lambda = \frac{h}{mv_n} \quad \text{with wave-particle duality, to give} \tag{16.2}$$

$$v_n = \frac{nh}{2\pi m r_n} \tag{16.3}$$

Coulomb's Law also provides a relationship between r and v, because the Coulomb force holds the electron in circular motion:

$$\frac{mv_n^2}{r} = k\frac{e^2}{r_n^2} \quad \text{and combining with 16.3} \tag{16.4}$$

$$r_n = \frac{h^2 n^2}{4\pi^2 k q^2 m} \tag{16.5}$$

1. Find the radii of the first 5 inner orbits of the electron in the hydrogen atom.

16.2. OTHER RESONANCE EXAMPLES

Energy levels of hydrogen: 2. Fill in the following table for the lowest 5 energy levels of hydrogen. Use the values of r_n found in question 1) above. Use equation 16.3 to find the corresponding v_n's. Find the KE using the usual formula for the kinetic energy of a particle of mass m, and the PE using $PE = -ke^2/r$. Final column is TE = KE + PE.

n	r_n	v_n	KE	PE	TE
1					
2					
3					
4					
5					

3. The bright lines in the hydrogen spectrum occur when the electron drops from a higher orbit to a lower orbit. *The energy difference, when divided by Planck's constant, gives the frequency of the emitted light.* Lines must fall in the wavelength range, [400 nm , 700 nm], in order to be visible.

All transitions between n = 2, 3, 4, ... and n = 1 fall below 400 nm, in the ultraviolet. They are not visible. Transitions falling from higher levels down to n = 2 emit light in the visible wavelength range.

Find the wavelengths for the following transitions: $5 \to 2$, $4 \to 2$, and $3 \to 2$.

Answers: 2. TE(n = 1) = 2.17×10^{-18} J, or 13.6 eV   3. 656 nm (red), 486 nm (blue-green), 434 nm (blue)

Chapter 17

Lens optics

Light travels in straight lines until it reaches the interface between *different* materials. Section 14.3 discussed refraction, or bending of light at an interface.

Two interfaces: 1. A light beam enters a glass plate at a 45° angle, as shown in figures 17.1 a) and b).

a. What is the angle of refraction (the refractive index of glass is 1.50)?

b. In figure 17.1 a), extend the refracted ray past the righthand glass/air interface, and show that the final refracted ray leaving the glass is parallel to the incident ray on the lefthand air/glass interface.

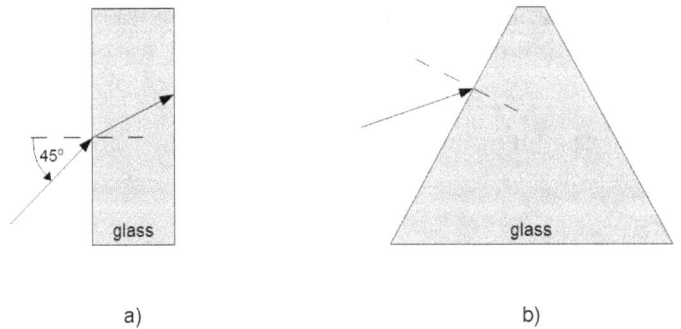

Figure 17.1: Light rays traversing glass slabs

c. In figure 17.1 b), trace the incident ray through the trapezoidal glass slab, and show that the ray emerging from the righthand interface is sharply bent down from its initial angle.

Answers: 1. $\theta_r = 28.1°$

2. a. Figure 17.2 a) shows three parallel light rays striking three chunks of glass. Trace each ray through the glass. You do not have to calculate the angles of refraction, but you should use the following basic principles:

- When a ray passes into a denser medium – e.g., from air to glass – the ray will bend *toward* the normal.

- When a ray passes out of a dense medium – e.g., from glass to air – the ray will bend *away from* the normal.

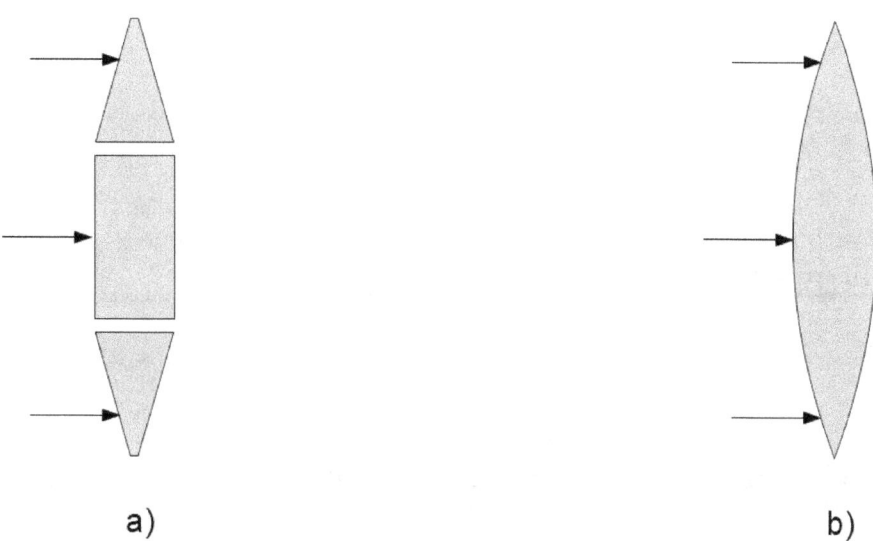

Figure 17.2: a) Three separate chunks of glass ; b) glass lens

b. In figure 17.2 b), draw the three outgoing rays from a lens.

17.1 Focal length, source and image distances

The focal length of a lens is defined as the distance from the lens to the point where parallel rays converge. See Figure 17.3:

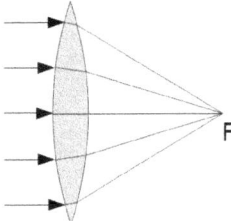

Figure 17.3: Parallel rays converge to a point a distance F from the lens

1. You use a lens with focal length 15 cm to burn a hole in a piece of paper. How far from the lens do you hold the paper?

If a source of light is a distance S from a lens with focal length F, the image distance I is given by

$$\frac{1}{I} = \frac{1}{F} - \frac{1}{S} \quad \text{or}$$
$$\frac{1}{F} = \frac{1}{S} + \frac{1}{I} \tag{17.1}$$

2. You use a lens with focal length 15 cm to image a ceiling light on a piece of paper. The light is 300 cm from the lens. How far from the lens do you hold the paper?

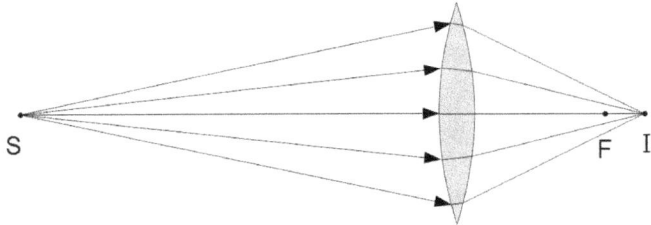

Figure 17.4: Rays from a distant source converge to a point just beyond F

Answers: 1. 15 cm   2. 15.8 cm

17.1. FOCAL LENGTH, SOURCE AND IMAGE DISTANCES

3. You use a lens with focal length 15 cm to image a small light on the desk up onto the ceiling. The light is 15.8 cm from the lens. How far from the lens is the image on the ceiling?

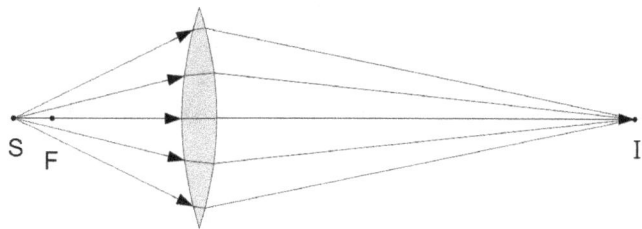

Figure 17.5: Rays from a source just outside the focus converge to a point far from the lens

4. If the focal length is 15 cm, and the source distance is 30 cm, what is the image distance? (See figure 17.6)

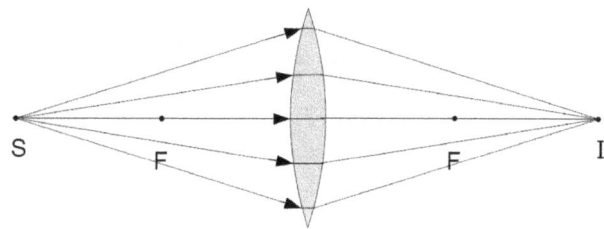

Figure 17.6: If the source is at 2*F, the image is also at 2*F

Answers: 3. 296 cm   4. 30 cm

17.2 Ray tracing and image size

In order to determine the image size, it is necessary to trace rays coming from the top and bottom of the source, and find out where they land on the image. We have already seen, in Figures 17.5, 17.4, etc., that source points on the axis of the lens wind up as image points that are also on the axis. We need also to trace light rays from a point that is off the lens axis. This is illustrated in Figure 17.7:

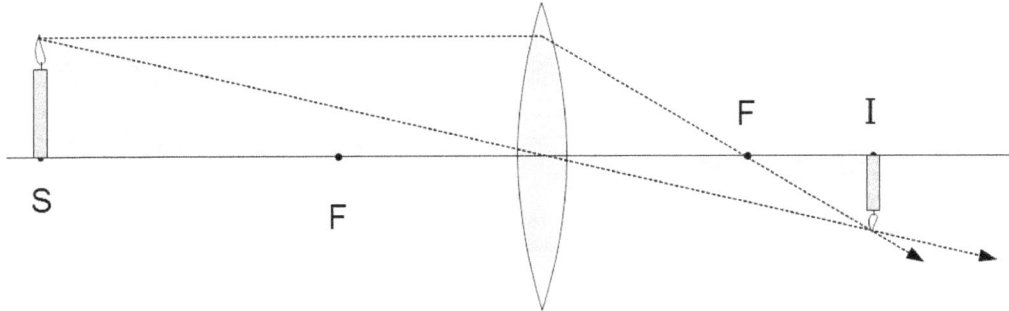

Figure 17.7: Two rays emitted by the candle flame at left are traced until they cross on the right. **Where they cross locates the candle image**

Proportional sides of the shaded, similar triangles in Figure 17.7 result in the following equation for image size:

$$\frac{size_I}{size_S} = \frac{I}{S} \qquad \text{Image size equation} \qquad (17.2)$$

1. a. Characterize the image in Figure 17.7.

Answer: The image is **inverted** and **reduced** in size, compared to the source. Traced rays cross through and are emitted in many directions from the candle *image*; therefore the image is **real**.

b. In Figure 17.7, focal length F = 12 cm, and the candle is 30 cm from the lens. What is the location of the image?

c. If the height of the candle is 8 cm, what is the height of the candle image?

17.2. RAY TRACING AND IMAGE SIZE

1. There is a candle on the table a distance 15 cm from a lens. The candle is 12 cm tall. The focal length of the lens is 10 cm.

   a. What is the distance of the candle image from the lens?

   b. What is the size of the candle's image?

   c. Draw a ray diagram for the source - lens - image using Geometrical Optics.

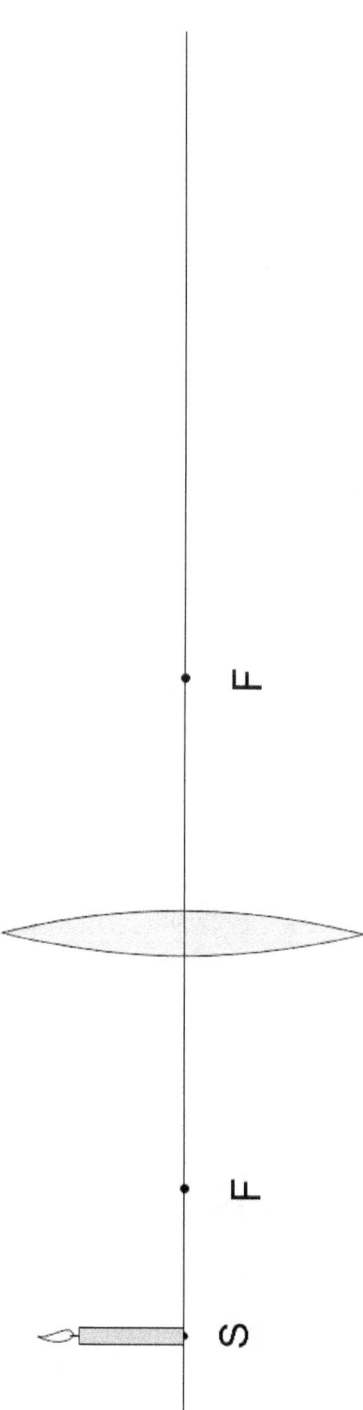

The image is (circle the correct ones) : (upright   real   virtual   inverted   magnified   de-magnified)

Answers: 1. a. 30 cm   b. 24 cm

2. There is a candle on the table a distance 20 cm from a lens. The candle is 12 cm tall. The focal length of the lens is 10 cm.

a. What is the distance of the candle image from the lens?

b. What is the size of the candle's image?

c. Draw a ray diagram for the source - lens - image using Geometrical Optics.

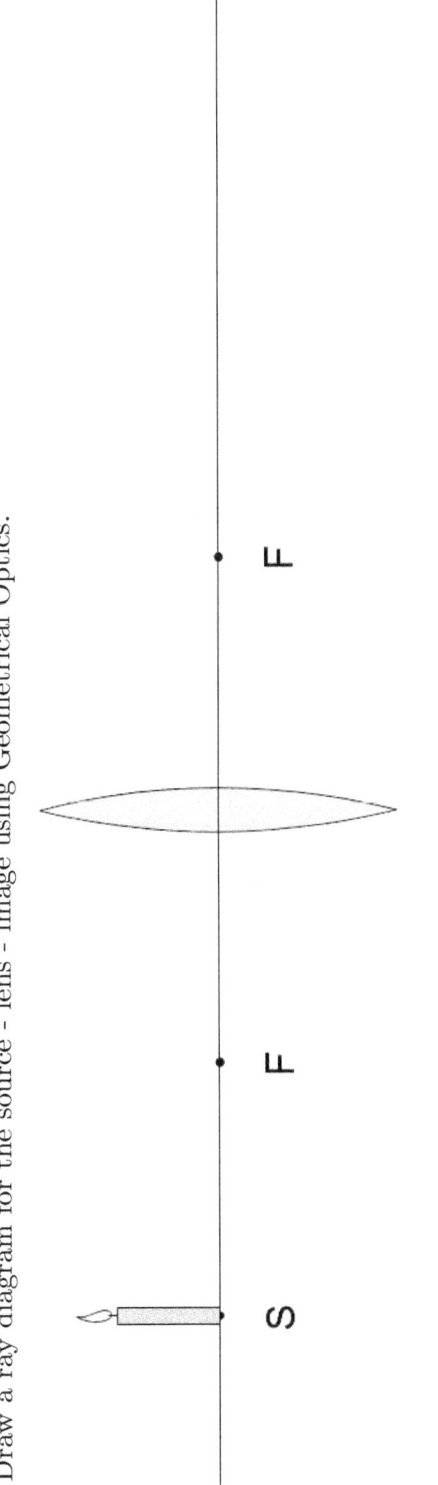

The image is (circle the correct ones) : ( ) upright   real   virtual   inverted   magnified   de-magnified ( )

Answers: 2. a.   20 cm   b.   12 cm

17.2. RAY TRACING AND IMAGE SIZE

3. There is a candle on the table a distance 30 cm from a lens. The candle is 12 cm tall. The focal length of the lens is 10 cm.

a. What is the distance of the candle image from the lens?

b. What is the size of the candle's image?

c. Draw a ray diagram for the source - lens - image using Geometrical Optics.

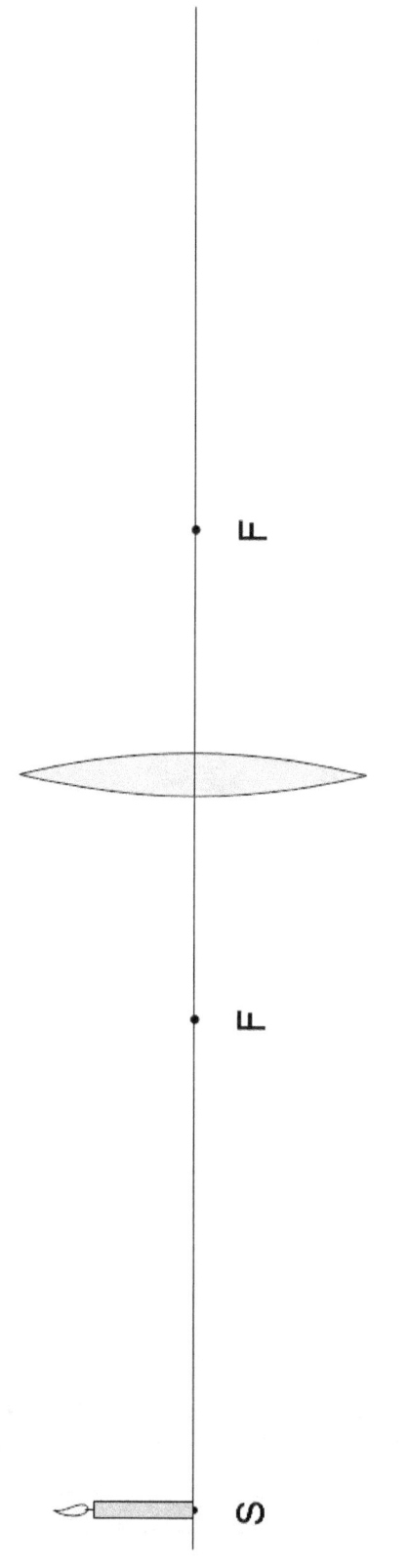

The image is (circle the correct ones) : (upright   real   virtual   inverted   magnified   de-magnified)

Answers: 3. a. 15 cm   b. 6 cm

17.3 Magnified virtual images

Problems so far in section 17.1 showed image formation that was real and inverted. When the source is closer to the lens than the focal length, $0 < S < F$, the ray trace becomes very different:

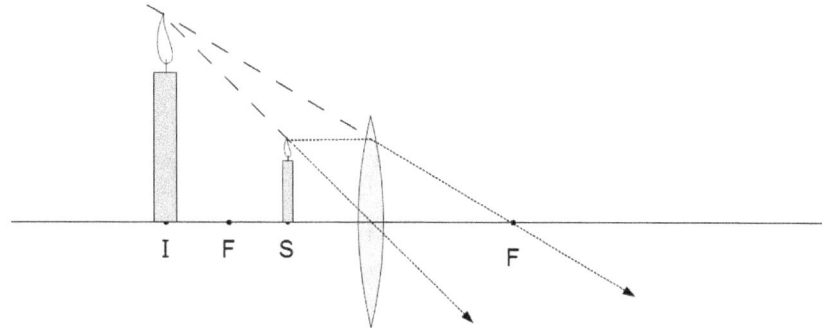

Figure 17.8: The light rays diverge and give no image toward the right, but can be traced backwards to a virtual image on the left.

The light rays traced in Figure 17.8 do not converge to an image point to the right of the lens. Instead, they can be traced backwards to an enlarged, upright *virtual* image to the left of the lens. The image is called 'virtual' because light rays do not actually pass through the image. They just *appear* to emanate from the virtual image location.

1. A lens with a focal length of 10 cm is placed 6 cm from a 0.5 cm object under study.

a. What is the image distance?

Answer: a. Using equation (17.1), the image distance is $I = 1/(1/10 - 1/6) = -15$ cm. The image is 15 cm from the lens. The $(-)$ sign indicates the location is not on the right of the lens, but rather to the left of the lens.

b. Using equation (17.2), the image size is $size_I = 0.5$ cm $\times$ 15 cm/10 cm $= 0.75$ cm. Here we have taken the absolute value of S; if we use the actual negative value, the image size calculation is negative. We would then interpret this to mean that the image is non-inverted.

2. A lens has focal length 10 cm. What are the image location and size if the source is placed 8 cm from the lens?

Answers: 2. image 40 cm behind the lens, magnified 4×

1. Consider a convex lens with focal length F = 5.0 cm. This will be used as a magnifier. The source is a small bug whose size is $size_S = 0.2$ cm. The bug is placed at three different distances from the lens: a) S = 2.5 cm; b) then S = 4.0 cm; c) then S = 4.5 cm. In each case, draw the ray diagram using geometric optics, calculate the image distance and image size.

a.

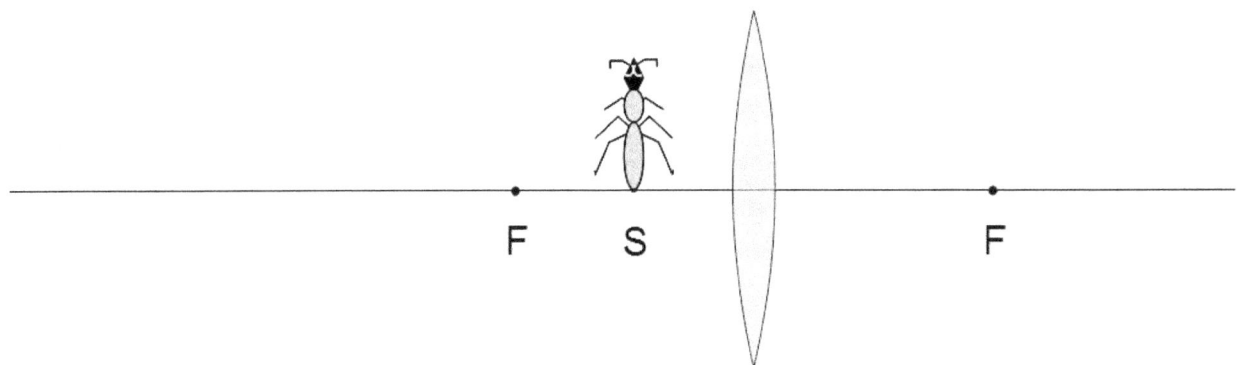

Figure 17.9: a) bug is at S = 2.5 cm from lens

I =

$size_I =$

b.

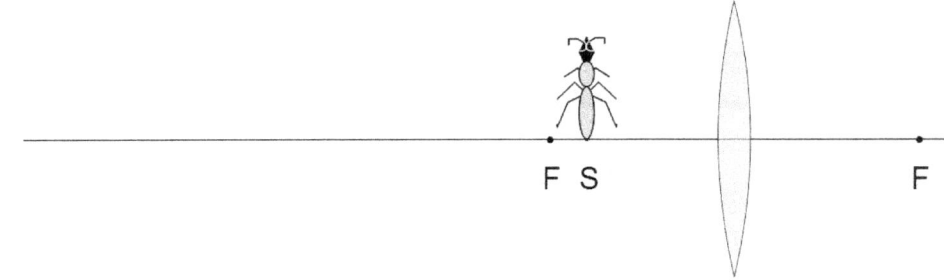

Figure 17.10: a) bug is at S = 4.0 cm from lens

I =

$size_I =$

17.3. MAGNIFIED VIRTUAL IMAGES

c.

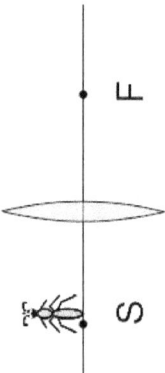

Figure 17.11: a) bug is at S = 4.5 cm from lens

I =

size$_I$ =

www.ingramcontent.com/pod-product-compliance
Lightning Source LLC
Chambersburg PA
CBHW082330220526
45470CB00008B/2460